YOUR SOUND ONSTAGE

YOUR SOUND ONSTAGE

Emile D. Menasché

Distributed by Hal Leonard Corporation

Published in 2011 by In Tune Partners
582 North Broadway
White Plains , NY 10603

Distributed by Hal Leonard Corporation
7777 West Bluemound Road
Milwaukee, WI 53213

Cover Photos: Digital Vision/Thinkstock; Background: Hemera/Thinkstock;
Author Photo: Monica Zane. Every reasonable effort has been made to contact copyright
holders and secure permissions. Omissions can be remedied in future editions.

Printed in the United States of America

Editorial Director: Emile D. Menasché
Book Design: Jackie Jordan

Library of Congress Cataloging-in-Publication Data is available upon request.

ISBN: 978-1-61774-231-6

intunemonthly.com

Dedication

For John Montalto, Doug Lane, and Bob Hoffmann—old high-school bandmates who still play a part in my music—and Steve Powers, drummer, saxophonist, and mentor in many things music and media.

And to the memory of Jay Junker, bassist and all-around great guy who was always there to carry an amp or two.

Contents

Chapter 1
The Elements of Sound Reinforcement

Chapter 2
Mixers

Chapter 3
Amplifiers and Speakers 31

Chapter 4

Chapter 5

Chapter 6

Chapter 7

Chapter 8

Chapter 9
Setting Up Your Sound System

Chapter 10

Setups for Microphones and Pickups

Chapter 11

Chapter 12

Chapter 13

Live Sound and Hearing Safety

Chapter 14

Appendix 1

Appendix 2

Appendix 3

Bonus

On the Disc

Video

THE ANATOMY OF A SOUNDCHECK

MIKING A SPEAKER CABINET

USING A PARAMETRIC EQ

USING A HI-PASS FILTER

USING A LO-PASS FILTER

USING A NOISE GATE

SOFTWARE INSTRUMENTS EXPLAINED

SAFE HEARING

IN TUNE COMMERCIAL

Audio

GAIN EFFECTS

INPUT LEVEL SETTING

MODULATION EFFECTS

DRUM MIKING EXAMPLES

GUITAR MIKING EXAMPLES

Resources

LINKS

PRINTABLE FREQUENCY CHART

PRINTABLE GLOSSARY

Foreword

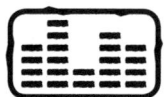

A SOUND CAN CHANGE YOUR LIFE

’ll never forget the day rock & roll changed my life. I was 15, and I had driven my ’57 Chevy to Laurel, Miss., to see Bo Diddley perform. Bo took the stage with his guitar and stood in front of an amplifier that had about 12 speakers in it—just an enormous amount of sound—and he let loose with the most outrageous guitar licks I’ve ever heard. Everyone went nuts. I just stood there watching his fingers move around the fretboard, watching his feet move with the beat, and feeling the power of that music as it pumped from the speakers. I was never the same again.

I went home after that show and told my folks I wanted to be a guitar player. My father, who owned the local music store in town, was adamant that I learned how to play on an acoustic guitar before he would consider giving me an electric. Well, there are many nice things you can do on an acoustic guitar, but play like Bo isn’t one of them. I was impatient, and I knew a thing or two about electronics from reading *Popular Science* magazine. Over Christmas break from school that year, I wound my own guitar pickup and put steel strings on that acoustic guitar. Then I rummaged some old television parts, salvaged a speaker from a jukebox, got out my soldering iron and built the first Peavey amplifier.

The science of sound has come a long way since that week in 1957. The amplifiers and sound systems you use (and I make) are vastly more advanced in design and features, and they sound plenty better, too. Today you can plug into a single guitar amp that models the tones of two dozen other amps. You can design your own virtual amp in software, or even play guitar through your cell phone. It's an exciting time to play music because the possibilities are so vast. And the boundaries keep pushing further and further out.

Still, many of the same questions that bothered me in my early days continue to dog musicians of all stripes, from beginner to advanced. "How many power amps and speakers do I need?" "What is the difference between a 4-ohm cabinet and an 8-ohm cabinet?" "How do I hook up all of this gear without blowing it up?" This book does a good job of explaining all of these things I had to learn the hard way.

Creating good sound for your band is about more than being loud. Sometimes volume isn't the issue at all. It takes time, practice, and knowledge to know when to tweak the gain and when to boost the high or upper-mid frequencies. It takes experience to know where your instrument should "sit" in the mix of instruments and performers in your band. Over time, these things become second nature … if you take the time to learn it all now.

I learned a lot of lessons by plugging in cables and cranking the knobs. Surely you've had a similar experience. But spend some time with this book, and pretty soon you'll have what it takes to make a garage band sound like they're playing Madison Square Garden. Here's a hint: It doesn't take 12 speakers to do it, either.

Good luck!

Hartley D. Peavey
Founder and CEO, Peavey Electronics Corporation
Peavey.com

Introduction

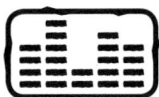

LEARNING IN STAGES—
AND ON STAGES

Like a lot of musicians, I really started to get into music when I was in high school. I'd been in orchestra and played at many a school concert, but my first taste of live performing that felt like a show was when the band my friends and I put together had our first gig. We'd been rehearsing at our parents' houses for a long time, and we were now about to go out and play for someone other than our folks, siblings, and pets (and, I imagine, angry neighbors).

There was only one problem. We had a few amps and instruments, but not much else. I used my parent's stereo for an amp at one point! Actually, there was another problem. We weren't getting booked into music clubs—we were too young—so our shows were at "venues" like backyard parties and dances in the school gym. We had to borrow and rent everything we needed. We had to set it up ourselves. We had no idea what we were doing. You might say we were just a little loud and out of balance. Okay, more than a little. You may still be able to hear our drummer's floor tom echoing around the JFK High School gym. To be honest, we didn't care that much. It was still fun to play.

But we never got invited back.

Fast-forward 6 or 7 years: My high school buddy John (bass) and I were on the stage of

New York's legendary alternative rock club, CBGB, along with our bandmates Alan (keyboards, lead vocals) and Mike (drums). Our band Club Iguana had been getting better and better time slots. We were about to do our first soundcheck as headliners.

We were all pumped, but also a little nervous. Not about the show—we played CB's pretty regularly—but because our regular sound man wasn't there. We relied on him! Not only that: As we walked in, the burly bouncer—the club was in a really tough neighborhood, so this guy was no joke!—told us that the house sound man was the only person in the world he feared.

Uh-oh.

Fortunately, we'd had a lot of practice setting up our gear from playing gigs in clubs and at parties all over the New York City area. We were efficient. Before every important gig, we'd rehearse our sets as if we were playing a show. We timed everything and knew how to move from song to song. We knew how to cover when someone (always me) broke a string. We had improvised sections, but we knew how to get in and out of them.

We worked on our sound at rehearsal and everyone understood what the other players' instruments would sound like on every song. We all knew our amp and effects settings down cold. More important, we thought of our sound as a "band thing." If someone didn't like the tone of my guitar, they spoke up, and vice versa. We worked together to create a blend.

We brought all that onto the stage that night, and as we started the check, the scary sound guy lost some of his bite. We patiently waited until he was ready for us to play. We asked him a few questions, but we did our best not to waste any time since he had five other bands to check after us. After he got our individual sounds, he brought the whole band together. He seemed to pick up on our approach quickly, so we were a little less nervous about how the night would go.

When we hit the stage a few hours later, our sound seemed to explode through the speakers. We weren't loud, but we had power—this guy knew how to take what we were doing to another level. That was one of the best shows we ever played together. Everyone was in the zone. I have no doubt that scary sound man's work helped us play better because we sounded better to ourselves.

The moral of this story: Great stage sound involves a lot of partnerships—between you and your bandmates; between the band and the sound engineer; most important, between performer and audience. In a way, that partnership even extends to you and your instrument and audio equipment. The better you know it, and the more you respect its role in helping you communicate, the better you'll sound onstage.

Communication is key. Had we asked, we might have learned that the bouncer was afraid of the sound man because the sound man was his older brother.

How to Use This Book

Your Sound Onstage is intended for young musicians (or at least musicians who haven't had a lot of stage experience), as well anyone interested in learning to set up and operate a basic sound reinforcement system. Think of it as a practical guide, not as a technical manual. The book is designed to give you an overview of the equipment and basic techniques used to project sound to an audience.

The "Your Sound" part of the title is important. A P.A. system may seem like a group of electronic tools, but I hope this book gives you an understanding of how to use microphones, mixers, amps, effects, and speakers to enhance your own personal sound. Whether you're playing Carnegie Hall or a two-car garage, the sound always starts with your singing and playing.

In addition to the more than 200 photos and drawings in this book, you'll find tip boxes scattered throughout. Many of these come from professional musicians, teachers, and sound engineers who generously shared their experience with the young musician/sound engineer in mind. Not that there aren't any technical terms or tricky concepts in the book. When they appear, I've tried to explain them as simply as possible without bogging you down with jargon. You won't find eight-page explanations of impedance here—though you will learn where you can. And by the time you've read the first few chapters, you'll be able to use impedance as well as words like signal, gain, pickup pattern, polarity, and many others in polite conversation.

You don't have to read the chapters in order. Many of *Your Sound*'s key concepts are discussed in several places. However, I do suggest you start with Chapter 1 before digging too deeply anywhere else. It offers a good overview of sound reinforcement and introduces some very important terms.

Each of the next six chapters explores one equipment category: mixers; amps and speakers; microphones and pickups; signal processors and effects; cables and connectors; and accessories. These chapters introduce the basic features and options within those groups.

From Chapter 8 on, we start to see how all the equipment fits together in practical situations. You'll learn how to assess what you need for a show, communicate with venue staff, set up your individual equipment, and integrate it into the P.A. (or bypass the P.A. entirely!).

You'll also learn how to keep all your cables and connections organized, hear yourself on-stage, mic your voice and instruments, take control of your personal sound, get the most out of soundcheck, and start creating a basic mix. You'll even learn how to pack up your stuff for the show!

Perhaps the most important chapter in this book is Chapter 13, which explains how to protect your hearing. As a veteran musician who suffers from hearing loss, this one is especially dear to my heart.

The included CD-ROM contains an important video about this subject, along with some useful printable worksheets, checklists, and charts. These, along with the glossary and other resources found in the appendices, should help you organize and set up your sound. You can also find links and additional resources at media.intunemonthly.com/yso.

Finally, I want to thank you for taking the time to read this book. I hope it helps you sound better, have more fun making music, and connect with your audience, whether it's a huge crowd or simply a close friend.

—*Emile D. Menasché, White Plains, New York, May 2011*

Acknowledgments

Many people helped make this book possible. Special thanks to Hartley Peavey—the C.E.O. of Peavey Electronics—for writing such an insightful and inspiring foreword. Peavey's media relations go-to guy Jim Beaugez helped us track down images and wiring diagrams. Etymotic's Gail Gudmundsen supplied an important video about safe hearing that every musician and sound engineer should watch, regardless of age and experience.

Shure Incorporated's Davida Rochman, Gino Sigismondi, and Matt Engstrom provided excellent photographs and technical advice. Gino and Matt—along with audio engineers Phillip Jordan, Eric Turqman, Kevin Madigan, recording artist Amanda Shires, and musician/educator Charlie Lagond—deserve special mention for their Pro Tips scattered throughout this book. Charlie gets extra thanks for inviting me to the Lagond Music School in Elmsford, NY, and sharing his insight on teaching young musicians not only how to play well but also how to look and sound good onstage. We took many of this book's photos at an LMS concert, where I watched students put together the sound system—then use it to deliver a knockout show. Likewise, thanks to recording artists the Afters for showing us their professional soundcheck.

David Fish, Dean of Catawba College's popular music program, shared his experiences teaching popular music and technology. Kimberly Drake and Jennifer Paisley from NAMM helped us find some inspiring pictures from the 2011 SchoolJamUSA Finals. Phil Garfinkel, Steve Lipman, Diane Gershuny, and Shore Fire Media helped arrange interviews. Websites dpamicrophones.com, jbl.com, mackie.com, and rane.com provided lots of technical information. Rich Tozzoli and Jon Chappell provided useful advice. Mike Levine saved our CD.

Creative Director Jackie Jordan went way beyond the call of duty to make sure that every image appeared where it needed to be, often retaking murky photographs and redoing diagrams to make them both prettier and more useful. Mac Randall brought his combination of editorial skill, musical insight, and technical knowledge to this very challenging subject. He clarified my ideas—but also made sure I didn't forget about fuzz boxes. Proofreader Susan Kornfeld's eagle eyes caught a few embarrassing typos too. Production manager Robin Garber helped with layouts and illustrations. Young musician and editorial intern Ben Sparks also pitched in by gathering hundreds of gear photos—and posing in a few himself.

Your Sound Onstage wouldn't exist without the vision of In Tune Partners C.E.O. Irwin Kornfeld, who has reinvented the way the media talk to young musicians.

Chapter 1

THE ELEMENTS OF SOUND REINFORCEMENT

Sound reinforcement: It sounds so serious, doesn't it? Like something you might take into battle, or use to protect yourself from a hurricane. But that's why I like that term better than "P.A. system" when describing the equipment that's used to help musicians be heard onstage. A P.A. system—short for "public address system"—seems like something designed for a speech. Sound reinforcement, on the other hand, seems like something that will support the music you're making.

But whether you call it sound reinforcement, the P.A., or "that pile of stuff with the knobs and speakers," the key word is the last one: *system*. Many different parts are needed to capture instruments and voices and project them to an audience, whether you're performing in a small club, a church, the school auditorium, or the stage in a stadium.

In later chapters, we'll take a look at these individual components in more detail. But first, let's take a quick overview of what a sound reinforcement system actually does.

What's a Sound Reinforcement System?

Let's start with a definition: A sound reinforcement system is any collection of components that projects sound to a group of listeners. That's a very broad definition, and it encompasses many variations. You'll find portable systems designed for solo musicians that are not much bigger than a backpack, and huge systems with speakers stacked as high as a small building. You'll find systems with most of the components built into a single housing, and others that involve dozens, even hundreds of individual pieces of gear. But whether they're big or small, all sound reinforcement systems work by converting sound waves—which are produced when we play or sing, or make any other noise, for that matter—into electrical energy, called a *signal*. That signal is amplified and sent to loudspeakers, which convert it back into sounds we can hear.

How Sound Works

Sound waves move in *cycles*, and have two important properties. A wave's *frequency*—or number of cycles per second (measured in *Hertz*)—determines the pitch and tone it produces. Low-frequency waves produce low-pitched notes. High-pitched notes are produced by high-frequency waves. Low to high, human hearing ranges from about 20 Hz to 20,000 Hz (usually written as 20 kHz). You'll see frequency discussed over and over in relation to audio equipment. It will tell you how well a speaker can reproduce bass sounds, how effective a mic will sound on violin, or where to set an equalizer to sweeten a saxophone tone.

Amplitude describes the intensity of a wave as it moves through its cycle. It's measured in *decibels* (dB). Like frequency, decibels can be used to measure both the physical sound waves (*sound pressure levels*, or SPL) and the electronic signals that pass through audio equipment (called *gain*). Decibels are a little tricky because they follow a logarithmic scale and their measurement is relative to a given reference point. For example, 0 dB doesn't usually mean "silence." It's just a reference point that can be based on too many variables to discuss here. But you should know that an increase by 6dB represents a doubling in intensity: in other words, 6 dB is twice as loud as 0 dB, while -6 dB is half as loud as 0 dB.

When the gain of a signal at an input is the same as at its output, it's known as *unity*.

A BRIEF LOOK AT COMPONENTS

If you were to break a sound reinforcement system down to the absolute basics, there would be four main elements:

1. Microphones and/or pickups that turn sound waves into electrical signals.
2. A mixer, which is used to blend signals from mics and other sources.
3. An amplifier, which makes the blended signals from the mixer loud enough to cause a speaker to vibrate.
4. Loudspeakers, which project sound to the audience.

There's more to a sound reinforcement system than that. There will be the amps and instruments used by the performers. Most sound reinforcement systems also include devices called *signal processors* to help enhance the sound. We'll get to those later. But the essence of the sound reinforcement system is in those four basic components.

All of the things that make sounds—the instruments, voices, etc.—will enter the mixer

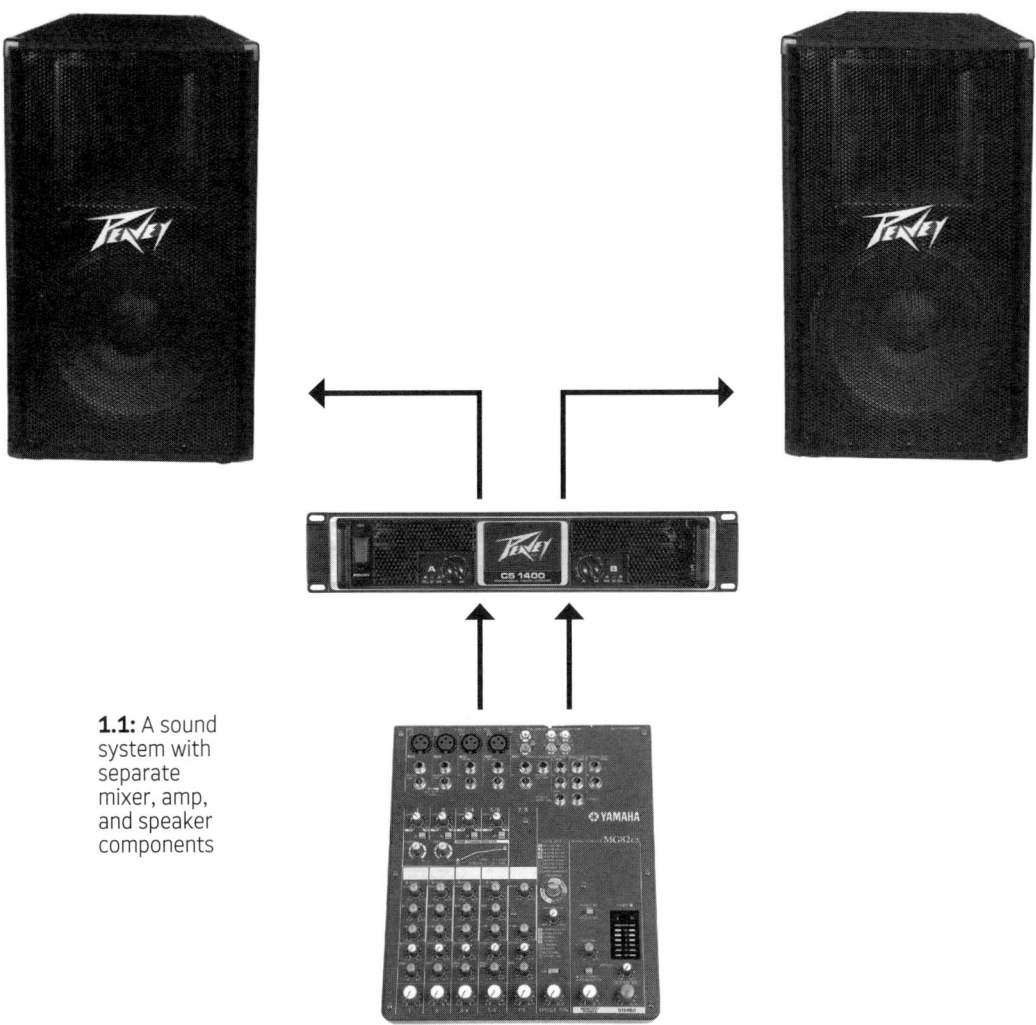

1.1: A sound system with separate mixer, amp, and speaker components

through the microphones and pickups plugged into the mixer's inputs. In turn, the mixer's outputs feed the amplifier, which then feeds the speakers.

Types of Sound Reinforcement Systems

In a *component system* (Figure 1.1), the mixer, power amps, and speakers are all separate devices. This offers a few advantages. First, it allows you to use the same mixer with larger and smaller sets of amps and speakers. It also makes it easier to solve problems when one component fails. If, for example, your amp breaks down, you can replace it without replacing the mixer or speakers.

The mixer, amp, and speakers can all be made by the same company or they can come from completely different manufacturers. You can also upgrade an individual component while keeping others—for example, replacing an old analog mixer with a new digital model and using that with the amps and speakers you already have.

On the other hand, component systems are more expensive (you have to buy more items) and they are harder to carry around and set up, simply because there are more individual parts to pack, unload, etc.—and more connections to make between them.

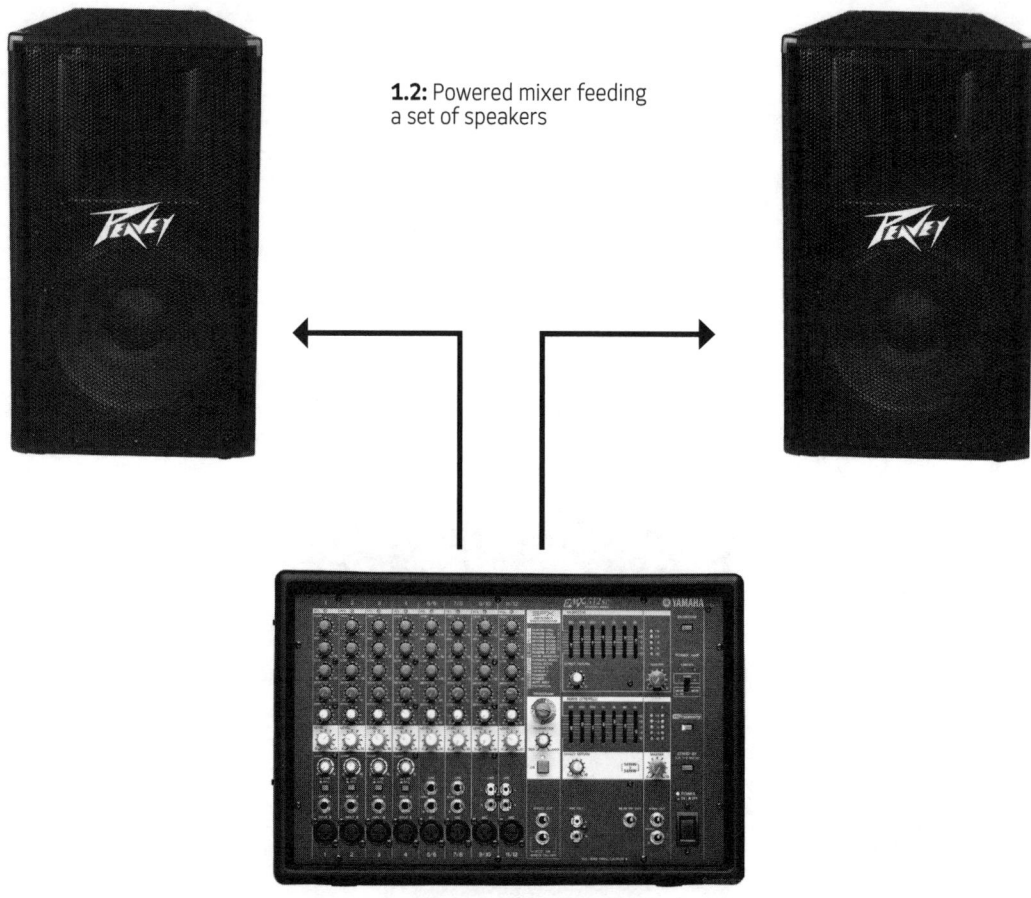

1.2: Powered mixer feeding a set of speakers

Combining Power With....

Professional sound reinforcement situations almost always call for a component system—some of them incredibly elaborate. But small ensembles often use systems that combine the power amp with another component: either the mixer (in what's known as a *powered mixer*) or the speakers (known as either *active* or *powered monitors*).

Figure 1.2 shows a powered mixer feeding a set of speaker cabinets. This saves a lot of setup time, but it also means that you can't easily change the amplifier or mixer if the situation calls for something different.

Figure 1.3 shows a standard mixer feeding powered monitors. Powered monitors are becoming more and more popular because they offer one distinct advantage over other systems. Since the amps are built into the speaker cabinets, they are designed from the very beginning to match the speakers inside. As we'll see in Chapter 3, this can help improve both the sound and reliability of your system. The disadvantage is that it's harder to add more power if the monitors aren't loud enough for your situation.

All-in-one systems take the "power-plus" idea even further. Here, you have a mixer, power amp and speakers all grouped together into one product. Figure 1.4 shows an example of an all-in-one system. These are great for solo performers because they're easy to carry around and set up. Some amplifiers designed for acoustic guitarists and keyboardists are

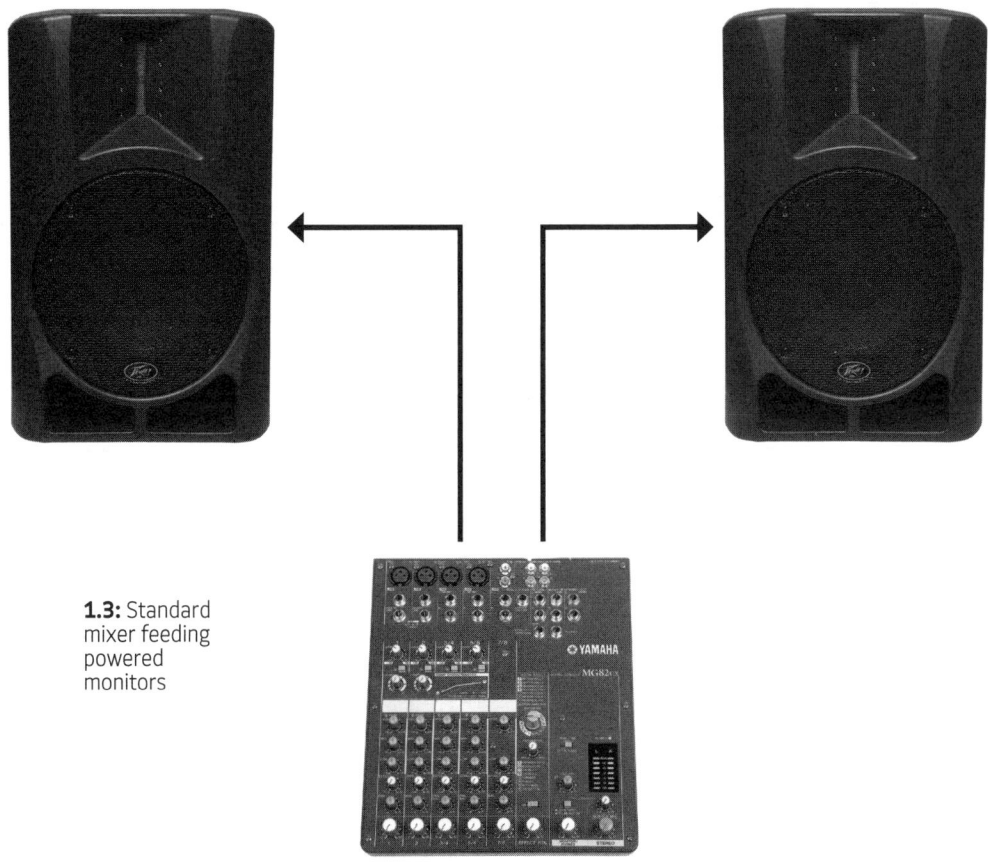

1.3: Standard mixer feeding powered monitors

1.4: All-in-one system

actually like miniature sound reinforcement systems, because they have the ability to mix more than one sound at a time. These instrument amplifiers are basically all-in-one sound reinforcement systems that are designed with a very specific user in mind.

Signal Path

Earlier, we explained that microphones and pickups convert sound energy into an electrical signal. This signal is like the blood in a sound reinforcement system's veins. It flows through your gear in what's known as a *signal path*. Every audio device in your system has a signal path. In addition, all the devices put together form a signal path. Put very simply, a signal path might look like this (Figure 1.5):

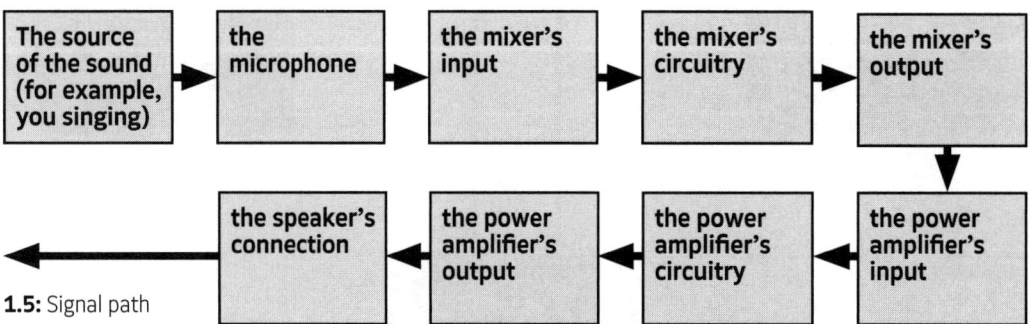

1.5: Signal path

Along the way, the signal's gain may increase or decrease, and its properties—such as frequency balance—may be altered many times over. But to simplify things, let's forget about what happens inside the mixer and amp for a moment and think of the signal like this: Plug the microphone into the mixer. Plug the mixer into the amp. Plug the amp into the speakers. Sounds simple, right? Actually, it is. Mostly.

On some systems, like the all-in-ones mentioned in the previous section, some of these connections are made inside the device. But in a component system, you need to plug everything in. And to do so, you must have the right kinds of cables.

Cables and Connections

We will be talking about making audio connections throughout this book and looking at them in detail in Chapter 6. But I want to take a second to make a point here: Cables may seem like "extras" when compared to items like mixers, amps, and speakers, but that's not the case.

First, it's important to have the right kind of cable to do the job—they're *not* all the same. Mixers, amps, and speakers all have very specific requirements. It's not rocket science to have the right cable, but you should be aware of the differences.

Second, quality actually matters here as much as anywhere. Nothing can mess up your sound as easily as a bad cable. I've known people who thought their amps were broken when the real problem was the wire running from the mixer to the amp.

A good audio cable has two or three wires enclosed within a protective covering, called the *casing*. That's the rubbery part on the outside. Between the rubbery outer coating and the important wires inside, there may be a layer of metal called the *shield*. This can help reduce noise. Audio cables come in two main kinds: balanced and unbalanced.

Power and Safety

Even the most basic sound reinforcement system requires a healthy amount of electricity to run. How you plug into the wall can affect sound, and more important, your safety.

■ Always use the electrical cables that came with your gear, or cables of the same or heavier gauges.

■ Never use a an electrical cable that's broken or frayed. Make sure the ground tip is attached!

■ If you use an extension cord, get a good heavy one. Avoid those thin cords from the dollar store like the plague.

■ Never allow electric cables to come into contact with liquid.

■ Get a good, heavy-duty surge suppressor and plug everything into it.

■ If possible, plug all audio devices into the same circuit—as long as the circuit is rated to handle the load created by all the gear: your mixer, amps, powered monitor, preamp effects, etc.

■ Anything that gets plugged directly into the mixer—for example, an electronic keyboard, or a bass amp with a direct out—should also be plugged into this circuit. This will prevent a ground loop, which can cause hum or worse.

■ Do your best to keep electrical cables away from audio cables, as they can introduce noise.

Balanced Cables

Balanced cables have three wires inside the casing: hot, cold, and ground. They're usually connected to one of two types of plug: an XLR, or a 1/4" tip/ring/sleeve (TRS). Figure 1.6 shows a balanced cable in raw form, and the different plugs it's designed for.

Balanced cables with XLRs are most commonly used for microphones, but they can also be used to connect mixers to amplifiers and to connect some preamps to mixers. Professionals usually prefer balanced cables whenever they can use them because they tend to sound less noisy and allow for longer cable runs.

1.6: Balanced cable with XLR plugs

1.7: Unbalanced cable with 1/4" TS plugs

Unbalanced Cables

Unbalanced cables have two wires inside the casing: hot and ground. They're usually connected to a 1/4" tip/sleeve (TS) plug or to an RCA plug. Sometimes, two unbalanced connections can be made with a single three-wire cable (the same kind of cable used for a single balanced connection). These are commonly used for stereo connection, and for connecting effects to mixers. Figure 1.7 shows the plug ends of an unbalanced cable.

Speaker Cables

Speaker cables may look very similar to unbalanced audio cables, but they don't use the same wiring. So while you might use an instrument cable to hook up a speaker in an emergency, it's not a good idea over the long haul.

Speaker wire is actually closer to electrical wire: it's heavy and usually isn't shielded. You can buy speaker wire in different *gauges*—thicknesses—and many people recommend that a longer connection between the amp and the speaker calls for a heavier gauge speaker wire. The two strands of cable are considered to be positive (+) and negative (-). Speaker wire can terminate in a 1/4" TS plug, bare wire, or special connectors. Figure 1.8 shows speaker wire and different connectors.

1.8: Speaker cable with various connectors

It's important that the +/- arrangement on the speakers agrees with that of the amp: that is, that the amp's positive output feeds the speaker's positive connection, and vice versa. If you connect more than one speaker, the way these terminals connect is even more important. We'll discuss that in detail in Chapter 3.

POLARITY AND PHASE

The whole +/- thing brings up one more important point about sound: *polarity*. Remember we mentioned that sound waves move in cycles? Well, those cycles go back and forth from a positive pole (+) to a negative pole (-), crossing a zero point in the middle (see the top graph of Figure 1.9). When two waves of the same frequency hit their positive and negative

Inside the Spec Sheet: Impedance

Here's another important technical term you'll see over and over: *impedance*. To impede something means to slow or stop its progress. This happens in electrical circuits, and it has an effect on the way audio flows through its signal path. I think *The Sound Reinforcement Handbook* by Gary Davis and Ralph Jones offers a pretty clear definition: "The total opposition to the flow of alternating current (read 'audio signal') in an electrical circuit." All inputs and outputs have an impedance rating, which is measured in *Ohms*.

This very technical definition doesn't necessarily mean much to musicians and sound engineers. So Davis and Jones point out a practical way to look at impedance: The impedance of an output defines how easily power will flow from that output. The impedance of an input measures how much power that input will draw. Ideally, the two should match or at least be close. As you'll see in Chapter 3, this is really important when you hook up amps and speakers. But it also makes a difference when you plug microphones and instruments into mixers and amps. The bottom line: You should always know the impedance of the inputs and outputs you're connecting. A device's *spec sheet*—which lists things like the device's frequency response, power rating, output level, and so on—will tell you what kind of impedance it likes to see. Do your best to follow those guidelines and you should be okay.

peaks at the same time, the waves are *in phase*. When one hits the positive peak at the same time that the other hits the negative peak, the waves are *out of phase*. This can cause them to cancel each other out, which can make them sound weak or thin. This can happen when, for example, two mics pick up the same sound from slightly different spots, or if a cable is miswired, or sound from two speakers are interfering with one another. You'll hear about frequency, gain, polarity, and phase many times in this book.

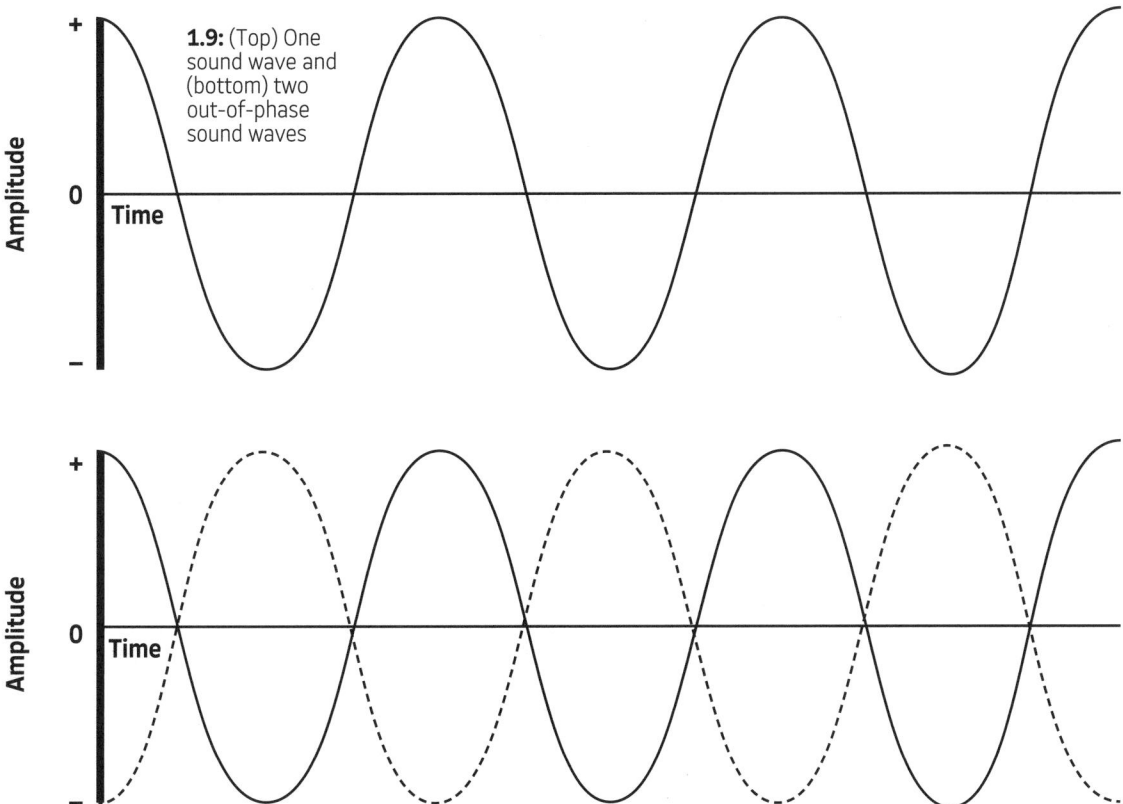

1.9: (Top) One sound wave and (bottom) two out-of-phase sound waves

MIC, LINE, AND INSTRUMENT LEVELS

Different analog audio connections have different voltage levels. These are measured once again in decibels, but relating to voltage (dBu or dBV). Microphones have the lowest; a low-impedance *mic level* is around -20dBu. (See page 8 for more about impedance.) Electronic devices like mixers have the highest; a low-impedance *balanced line level* is +4dBu. An *unbalanced line level* is lower than a balanced line level (-10dBV). Instruments like electric guitar and bass put out a high-impedance, unbalanced signal at a level generally somewhere between mic and line levels, sometimes referred to as *instrument level*. To plug an instrument into a balanced mic input, you need a *transformer*, usually found in a *direct box*.

Next Stages

Now that we've taken a look at the basics, let's get into some more detail about the individual components. We'll start with the most complex thing in your audio system: the mixer.

Chapter 2

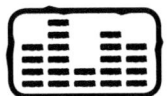

MIXERS

The mixer is the heart of any stage audio system. It's the component that combines all the sounds you and the rest of your ensemble make so that the audience hears one unified sound. But it can do even more than that! A mixer is like a railroad terminal for sound. If you want control of your sound onstage, it's essential that you have at least a basic understanding of how mixers work.

A mixer—also known as a *mixing console*, *mixing desk*, or *soundboard*—does what its name suggests: It blends sounds together and feeds that blend to an amp and speakers.

Unfortunately, if there's one piece of gear that scares people who don't have a lot of experience with audio equipment, it's the mixer. Look at all those knobs and buttons! Look at the sliders! Look at all the connections on the back!

Relax: A mixer isn't anywhere near as complicated as it looks. Mixers are arranged so that every connector goes through its own set of circuits, called a *bus* or *channel*. Typically, every channel has the same or very similar controls.

Once you learn how to use one mixer, you've pretty much learned them all—as long as you understand the basic operating principles involved in their design. You may need to add to that knowledge here and there to adapt to some variations in features, but that's not usually very difficult. Let's look at the basic features you might expect to find on various types of mixers, break mixers down into sections, and explain how all the inputs, controls, and signal paths fit together.

Types of Mixers

There are many different kinds of mixers, but they all do the job outlined in the preceding section. A mixer may be a huge console with dozens of channels or a tiny box with just a few. You'll find digital mixers—which convert the audio signal into digital data—and analog boards that use technology dating back to the early days of electronics. Some mixers can do a little of everything. Others specialize in live performance, DJing, or recording. There are *line mixers* that don't have inputs for microphones. You'll even find software mixers that are operated by computer programs and mixer sections inside of electronic instruments.

Figures 2.1-2.5 show a bunch of different mixers—and boy, do they look different! But their main purpose is the same. In that way, they're not unlike the kind of mixer you use in a kitchen. Those mixers take a bunch of ingredients and blend them in a way that—hopefully—tastes good. Audio mixers do the same thing with sound.

2.1: Mackie 1604-VLZ Pro analog mixer

2.2: Yamaha EMX512SC powered mixer

2.3: PreSonus StudioLive 16 digital mixer

2.4: Yamaha MG 3214 FX stage console

2.5: Pioneer DJM-700 DJ mixer

The Sections of a Mixer

In Chapter 1, we talked about signal path, also known as signal flow, and described how sound from a *source* (the word we use to describe mics, pickups, electronic instruments, etc.) can go into a mixer before feeding an amplifier and loudspeakers. Now let's break things down a little further and see how signals pass through a mixer.

Mixers have two main sections: the *input* section and the output, or *master*, section. Signal travels through the input to the outputs along what's known as a *bus*. A mixer can have only one bus or may have several, each sending the signal to a different destination.

THE INPUT SECTION

The input section is where you plug in microphones and instruments. It's also where you'll find the controls that let you blend these sources together and set their relative loudness—i.e., decide whether the keyboard or bass should be louder. Each input corresponds to a separate *channel*, and each channel has its own controls for setting its loudness and adjusting its tone—among other things. You can use the keyboard channel's tone control to make it sound a little brighter without affecting the sound of the bass.

2.6: An analog mixer's input section—each numbered row offers mic and line inputs for a single channel.

Figure 2.6 shows the connectors for the input section of a compact analog mixer. Notice that each channel has identical jacks, and that each one has different connectors for microphones and for other sources. Figure 2.7 shows the controls for the input section of that same mixer. As you can see, the controls are identical on all of the channels.

2.7: The controls for the input channels shown in Figure 2.6. Note the channel numbers at the top of the image. The faders at the bottom of each column control that individual channel's level in the mix.

THE MASTER SECTION

The master section is where the channels get blended together before they reach the mixer's outputs. Here's where you set the overall level of *all* the channels. You may also use the master section to blend in effects or change the tone of the entire mix. Raising the volume here makes both the keyboard *and* the bass louder; adding more treble here makes both the keyboard *and* the bass sound brighter, etc.

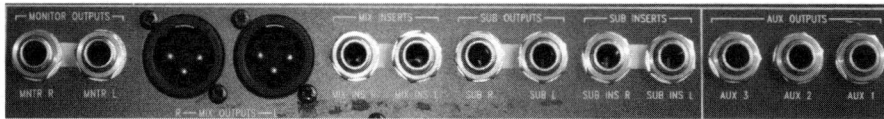

2.8: The master section of the mixer shown in Figures 2.6 and 2.7. Various connections (top) allow the mixer to route signal to several different destinations at the same time. The corresponding controls are shown at right. The master faders are at the lower right: These control the loudness of the complete mix.

Figure 2.8 shows the connectors and controls for the master section of a compact analog mixer. The main outputs are the ones typically used in sound reinforcement to feed the power amp (or powered monitors). You'll also find *auxiliary send* and *return* connections. These can be used to blend audio effects into the main mix, create alternative mixes (such the monitor mixes sent to musicians onstage), and more.

The master section lets you set the overall level of effects, the final output level of monitor mixes, and the level of the music going to the *house speakers* pointed at the audience. Many mixers are equipped with a headphones output with a volume control, and a separate monitor selector lets the headphones hear what's coming into the mixer even when the main outputs are muted. This is especially useful during setup and soundcheck. You may also find controls that switch the main mix with sound from another source such as an iPod (useful for playing house music in between sets).

Sub and Group Buses

A submix is like a mix within a mix. If your mixer has sub buses, you can use them to control a number of channels at the same time (we'll explain how to do this in the next part of this chapter). This can be especially useful when you want to control an instrument like the drum kit, which may have between three and eight mics, each on its own channel. By routing the drum mics to the sub bus, you can raise or lower the overall drum kit sound without disturbing the relative levels between instruments (Figure 2.9).

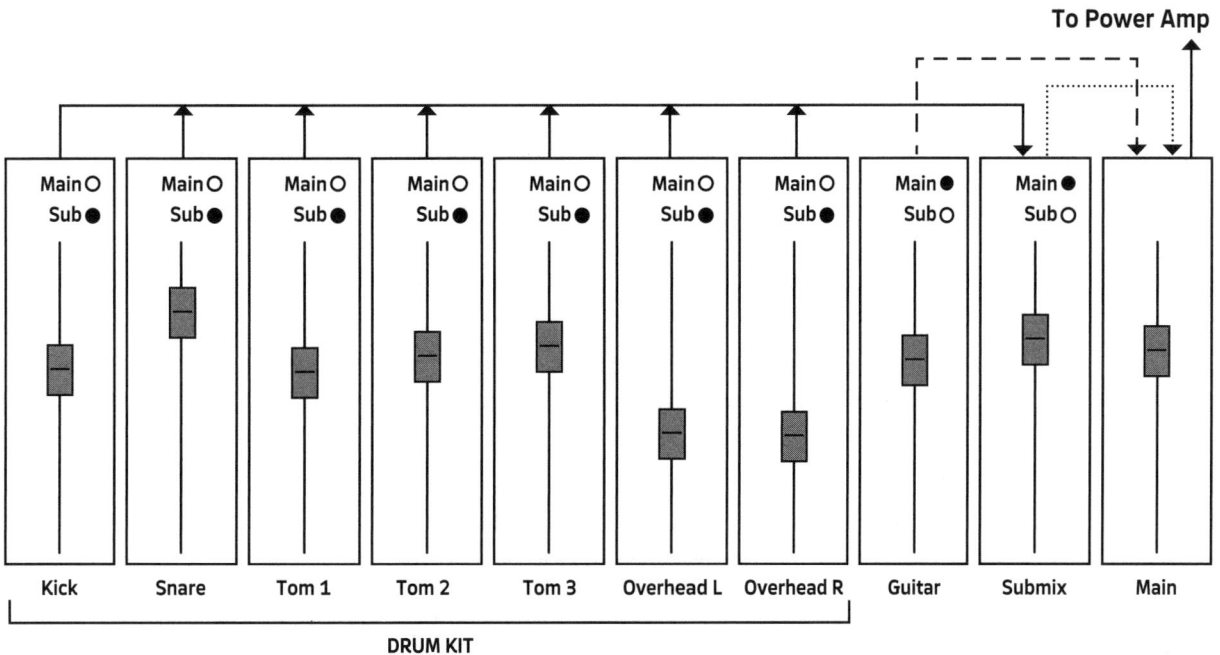

Fig 2.9: This illustration shows the faders from the input and master sections of a mixer. Drum mics are plugged into inputs 1-7 and a guitar is in input 8. The guitar feeds the master bus directly, but by routing the drum channels to a submix bus first, we can set the final level of all six drum channels with a single fader. The sub bus then feeds the main bus, which feeds the power amp and speakers.

Mixer Controls

Mixers don't just combine signal from different sources. They also let you adjust the sounds coming into each input so that they blend together nicely. A mixer will have controls that affect individual channels one at a time, which are grouped together into *channel strips*. It will also have controls that affect every channel at the same time. These are found in the master section.

You'll find lots of variation among different models. Some will have more bands of *EQ* (see next page), others will have a different number of sends, some may even have built-in effects. But the next section offers a general overview of what you're likely to find.

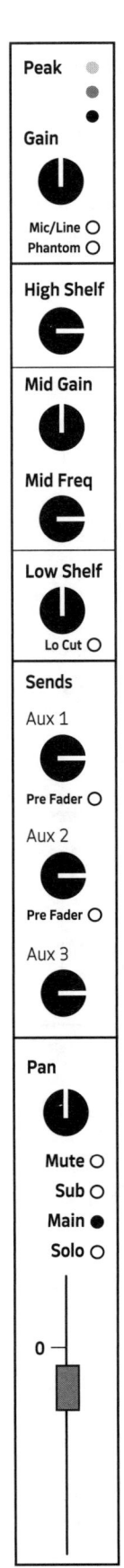

2.10:
Channel
strip

THE CHANNEL STRIP

If anything defines a mixer—other than the sheer number of inputs and outputs—it's the controls offered by the mixer's channel strips. Mixers usually have two kinds of input channels: mono (for handling a single channel of audio) and stereo (for handling two linked channels of audio, such as the outputs from an electronic keyboard).

We'll spend some time talking about how these controls interact when we get to mixing in Chapter 12. But for now, let's look at the basic elements of a channel strip.

Figure 2.10 shows a channel strip from a fictional mixer. The control layout is pretty typical of all mixers. Sometimes, these controls will be physical knobs. On digital mixers, some of these controls might be on menus inside the mixer's operating system. Either way, the signal will flow from top (the channel input) to bottom (the channel output).

Input Type and Phantom Power

If the mixer has separate connections for microphones and line-level signals on every channel, this control lets you choose which will be active. *Phantom power* is a special electrical signal that travels down a balanced microphone cable. It's used to provide the necessary power to some types of mics. (In some cases, phantom power can be used by other devices, such as acoustic guitar preamps.) Some mixers allow you to turn phantom power on and off on individual channels. Others have a global phantom power switch (affecting all mic channels) in their master section.

Gain

This is a crucial control: It sets the level of the signal coming into the channel. Your goal is to get a strong signal without creating an overload, or *clipping*.

Equalizer

The *equalizer*, or EQ for short, controls the balance between bass, midrange, and treble for each channel (we'll talk more about EQ in Chapter 5). Some mixers have very basic low and high tone controls that increase (*boost*) or reduce (*cut*) the sound at two preset frequencies. These EQs are often called *shelving equalizers* because their effect can be illustrated by a graph that looks like a shelf.

Instead of being fixed to bass or treble, more complex *parametric equalizers* can operate on the frequency range of your choice—be it bass, treble, or anywhere in between. Parametric EQs can be adjusted to focus very closely on one *center frequency*, or on a wider *bandwidth* ranging above and below the center frequency.

EQs are also known as *filters* because of the way they let only some parts of the sound come through. But wait—didn't we just say some EQs boost as well as cut?

That's true, but even when an EQ is boosting, it can't add what isn't already there. An EQ boosting very low bass notes will be useless on a violin, which is primarily a treble and midrange instrument.

Auxiliary and Effects Sends

Sends tap signal from a channel and route (or send) it to a destination that is independent of the main mix. On a mixer with built-in effects, one of the sends is usually hardwired to route a channel's signal to the mixer's onboard processor.

Sends are usually located after the EQ on a channel strip, so that the EQ you apply to the signal will also affect the sound coming from the send. Some mixers have a control that let you switch the send to *pre-* or *post-fader* operation. When set to pre-fader, the send's own level control determines how much signal will go into that send's output—no matter what you do with the channel's fader. Pre-fade sends are great for monitor mixes. They let you tailor the monitor mix(es) to suit the musicians' needs and feed that to the stage without altering—or being altered by—the mix the audience hears. Signal from a post-fader send changes in level as you move the channel's fader. This is useful for effects because the amount of signal sent to the effect will be in proportion to the channel's level in the mix.

Pan and Assign

In a stereo mix, *pan*—short for panorama—lets you control how a channel's signal flows between the mixer's left and right outputs. When pan is centered, the signal is equal on both the left and right. Rotate the pan control to change this balance. (Note, however, that in live situations, it's not uncommon to run everything in mono.) The *assign* buttons route signal from each channel to whatever buses are available—such as the main and sub buses discussed earlier. When a mixer has only one bus, there will be no assign button.

Mute and Solo

Mute turns the channel's output on or off; signal still comes into the input and can go to pre-fader sends. *Solo* lets you isolate one or more channels by muting all the others. Some mixers let you set the solo control to work in the monitor section of the mixer only, without interrupting the main mix feeding the house. This is an especially useful feature when making adjustments during a show, but not all mixers have it.

Fader

The *channel fader* is a slider that governs the overall level of the channel as it goes to the output buses. This is the control you'll probably use most while the music plays. Smaller mixers use knobs in place of faders. DJ mixers use a slider called a *crossfader*. Rather than control the level of a single channel, a crossfader sets the relative balance between two.

THE MASTER SECTION CONTROLS

All the signals in the mixer come together in the master section, which sets overall levels. Figure 2.11 shows the master section of our fictional analog mixer. The controls you'll find here are pretty typical of what you'll find in any standard audio mixer.

Aux and Effects Returns

These are used to add external signals to the mix, and to blend in the output of effects devices, such as reverb. If the mixer has built-in effects, one of the returns will control the overall level of the onboard processor as it mixes with the unaffected, or *dry*, signal coming from the channels into the master mix bus.

Equalizer

A graphic equalizer is a fairly common feature on sound reinforcement mixers. It can be used to tune the mixer to the room, though it must be said that EQing the overall mix can cause more problems than it solves. Either way, it's not an essential feature since you can always add an external EQ.

2.11: A mixer's master section may include level controls for send output and return input, monitor output, and headphones output. There may be a graphic EQ and faders for any sub buses. One thing you're sure to find, however, is a control for the main mix output (usually one fader or a pair, but sometimes a knob).

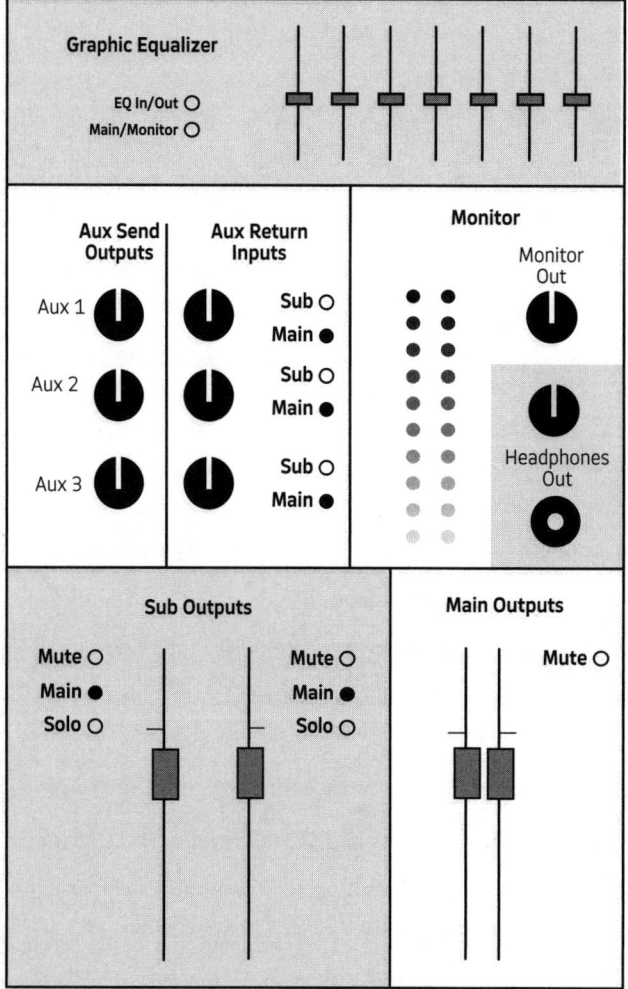

Monitor Section

Some mixers have controls that allow you to monitor independent of the main mix. We mentioned one application for this when we introduced the solo button in the previous section. If your mixer has a built-in headphones jack, you may be able to set it up to give the headphones their own independent mix. The monitor section will also include a level control and a set of buttons assigning it to hear the main mix, a sub mix, or another source.

Master Faders

Master faders set the level of the main outputs. Some mixers have independent outputs and controls for submix buses. In that case, the controls act like separate sets of master faders.

Signal Flow

The basic signal flow of a mixer goes like this: Sources come into the input section, where their individual levels are set. They're then routed to the output section. That's pretty simple, right? However, input channels can do more than route a signal; they can also affect the way it sounds. Just as important, on most mixers, they can send the signal to the main outputs and to other destinations—such as aux sends or sub buses—at the same time. On some mixers, individual channels can be set to bypass the main mix entirely!

If that seems a little confusing, it might help to realize that the way an audio mixer blends sounds is not all that different from the way a mixing bowl in the kitchen combines ingredients. But there's one big difference: In a mixing bowl, all the ingredients are lumped together, and once they're mixed, it's pretty hard to make adjustments.

With an audio mixer, all the individual ingredients remain separate and can be adjusted at any time. So instead of a mixing bowl, a better way to think of an audio mixer is as a set of pipes (Figure 2.12). In this illustration, each pipe feeds a different flavor of juice into a bigger pipe, which empties into a giant punch bowl.

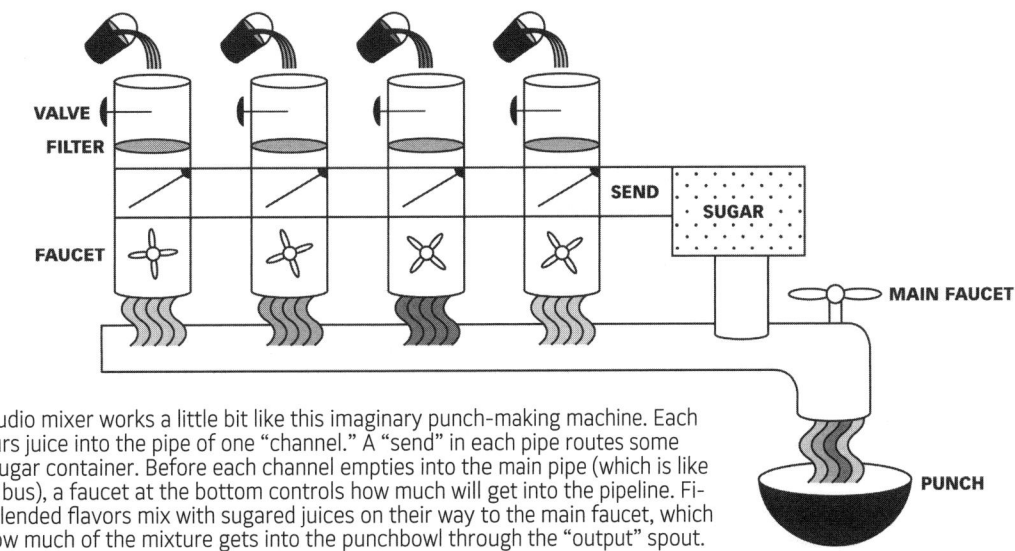

2.12: An audio mixer works a little bit like this imaginary punch-making machine. Each bucket pours juice into the pipe of one "channel." A "send" in each pipe routes some juice to a sugar container. Before each channel empties into the main pipe (which is like the output bus), a faucet at the bottom controls how much will get into the pipeline. Finally, the blended flavors mix with sugared juices on their way to the main faucet, which controls how much of the mixture gets into the punchbowl through the "output" spout.

Each of the feeder pipes has a funnel to let the juice in (on a mixer, this would be known as the *input*), and has a minimum of two controls. One determines how much juice gets from the funnel into the feeder pipe (on a mixer, this is *input gain*); the other—the *channel output*—determines how much of the juice inside the feeder pipe goes into the bigger pipe, which, on a mixer, would be an *output bus*. The big pipe also has a control called an *output*, which regulates how much punch goes into the bowl.

ONE INPUT, MANY OUTPUTS

While an input channel's main job is to send signal directly to the master section, most models also offer a way to route a channel's signal to a separate destination without changing the way it flows to the master section. These controls are called *assigns* and *sends* (they may be called either *effects sends or auxiliary sends*).

Assigning Channels to Buses

As we explained earlier, a channel's assign button determines where the mixer will send the signal at its output. Many small mixers have their channels permanently assigned to the master bus. But if your mixer has submix or group buses, you can use assigns to route each channel to the bus of your choice. For example, if you want channel 1 to go directly to the main outputs, press the "main mix" or "master" assign button. If you want it to go to a submix first, press that button (and then assign the submix to the main mix bus). You can usually assign a channel to the main and sub buses at the same time. If the mixer has output connections for the sub bus, you can route that to its own amplifier, audio recorder, or other destination independent of the main mix. Figure 2.13 shows how different assign settings affect the signal flow from the guitar channel.

What Size Mixer?

As you can see from the images in this chapter, mixers come in many shapes and sizes. So how do you know what size of mixer is big enough? The physical dimensions of a mixer do matter when you're using it for live performance—if you're stuffing your whole band's gear into the back of a Prius, you're not going to be able to take a seven-foot console to your gigs. But a more important way to measure a mixer is by the number of connections it offers.

Your mixer should have enough inputs for all of the sources you plan to route from the stage to the house system. That can include both microphones, instruments coming to the mixer via direct boxes, and sources coming to the mixer via line inputs. The mixer should also have enough independent outputs to suit your needs. If you plan to use auxiliary buses for monitors and effects, make sure the mixer has enough of them to cover you. Manufacturers have a standard way of describing a mixer's connections: They list the inputs and outputs as numbers, separated by the letter x. A mixer with 16 inputs and stereo outputs would be listed as 16x2. A 16-input mixer with four submix buses and a stereo output might be listed as 16x4x2. Auxiliary sends and returns are listed separately.

In addition to the number of inputs, you must consider their type. For eight microphones, you need at least eight microphone inputs. If an eight-channel mixer has four mic and four line inputs, it won't cover you. You can, however, feed a smaller mixer into a larger one and combine them into one big system.

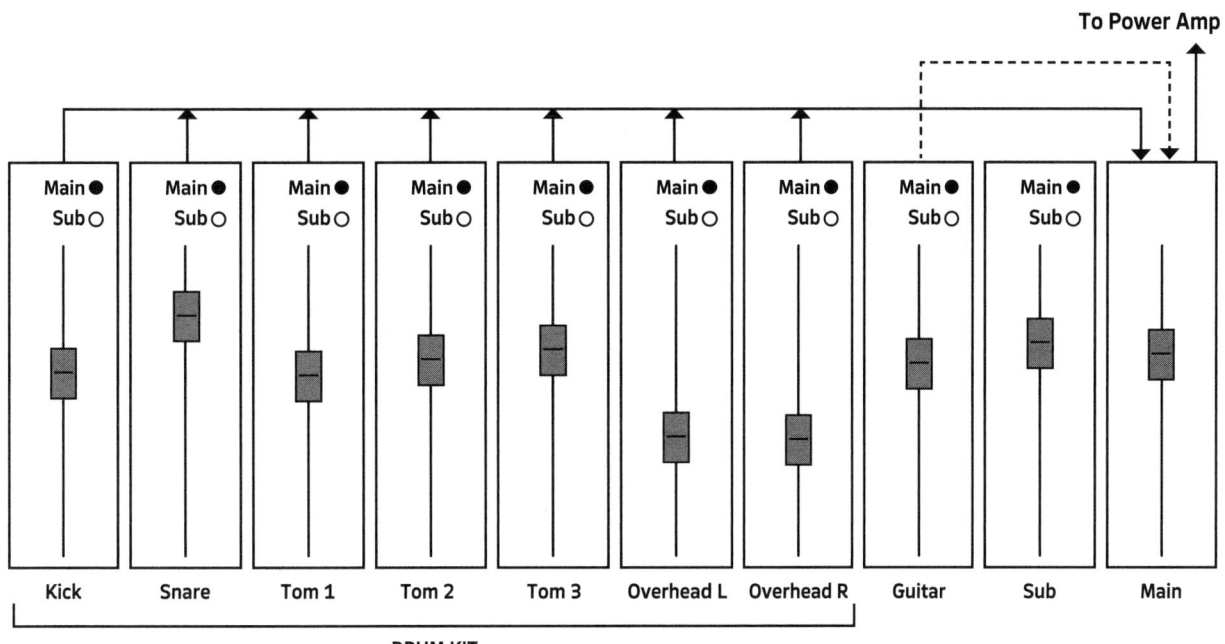

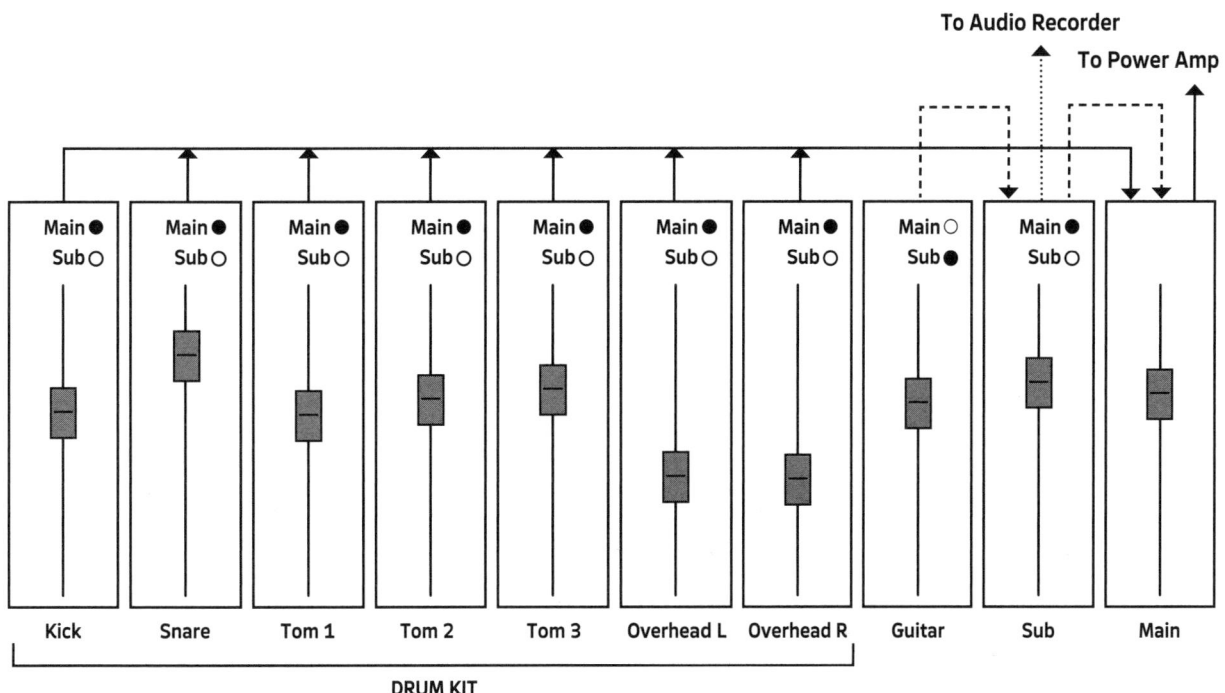

2.13: Using assigns, you can route the guitar channel directly to the main outputs (top) or send it to a submix bus first (bottom: Here, the submix is feeding both the main mix and an audio recorder offstage).

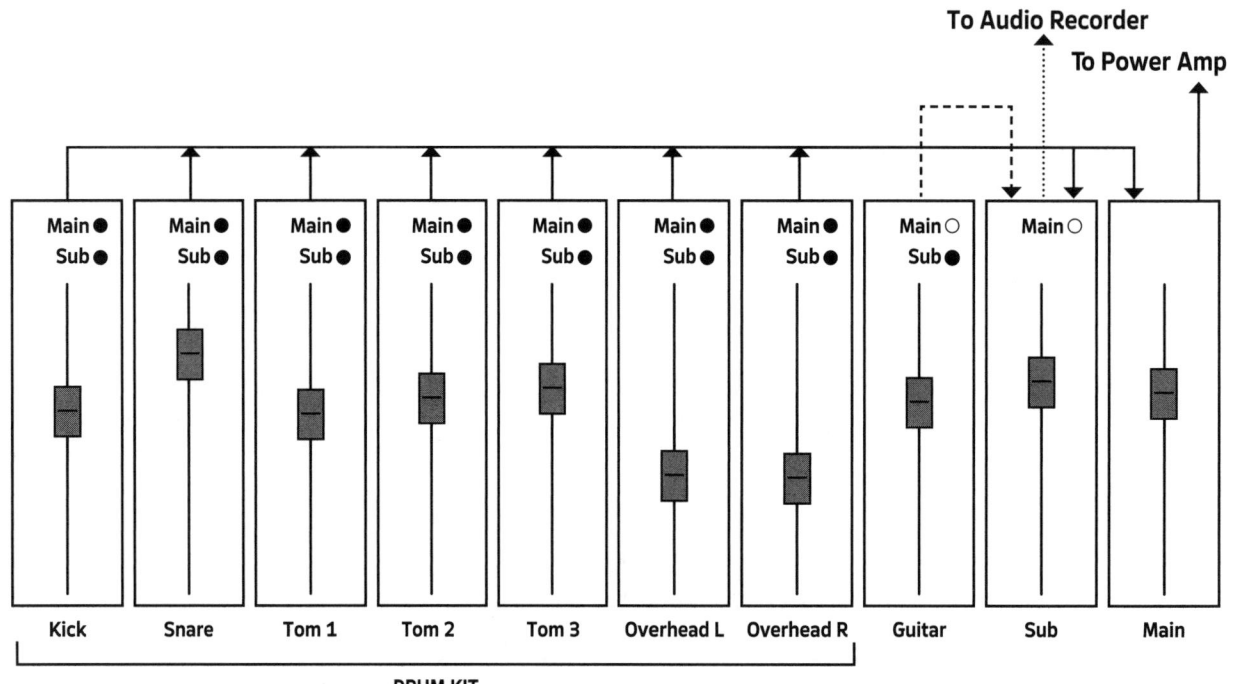

2.14: Here, the sub outputs are connected to an audio recorder, the main outputs to the house power amp. All drum channels are assigned to both the sub and the main buses. The guitar is assigned only to the sub bus. Although the the guitar is not in the main mix, the sub routes it to the recorder along with the drums.

You can use a sub mix bus as a "middleman" between one or more channels and the main mix—as we did with by grouping the drums in Figure 2.9. But a sub bus doesn't have to go to the main mix to be useful. Let's say you want to record your show by taking a feed from the mixing board, but because the guitar amp is already pretty loud onstage, you're not feeding it to the house speakers. A recording taken from the mixer's main outs would have plenty of vocals, drums, and bass—but not much guitar amp. Figure 2.14 shows how you could use a sub bus to include the guitar in a separate mix feeding the recorder, while keeping it out of the house system.

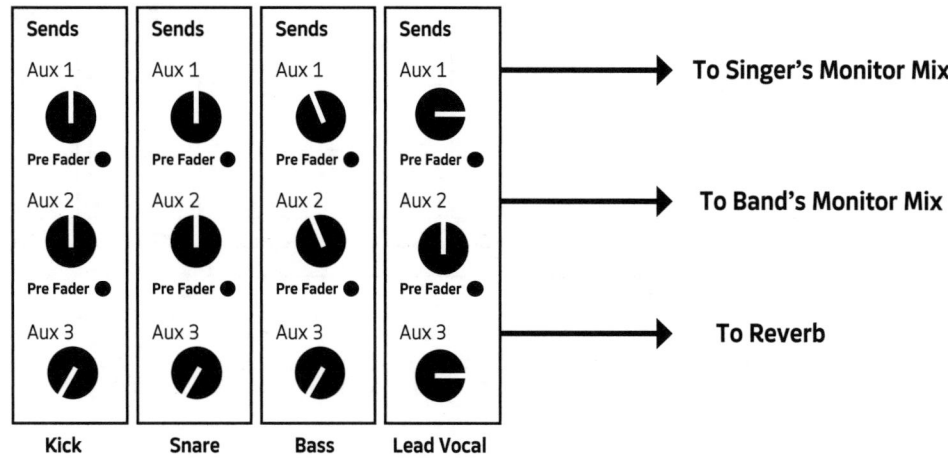

2.15: Aux 1 is feeding a singer's monitor mix. Aux 2 feeds a monitor mix for the rest of the band. Aux 3 sends the vocal channel a reverb effect. Note that the vocal level is higher in aux 1 (the singer's mix) than aux 2, and that both aux 1 and 2 are set to pre-fader so that the monitor levels won't change with the house mix.

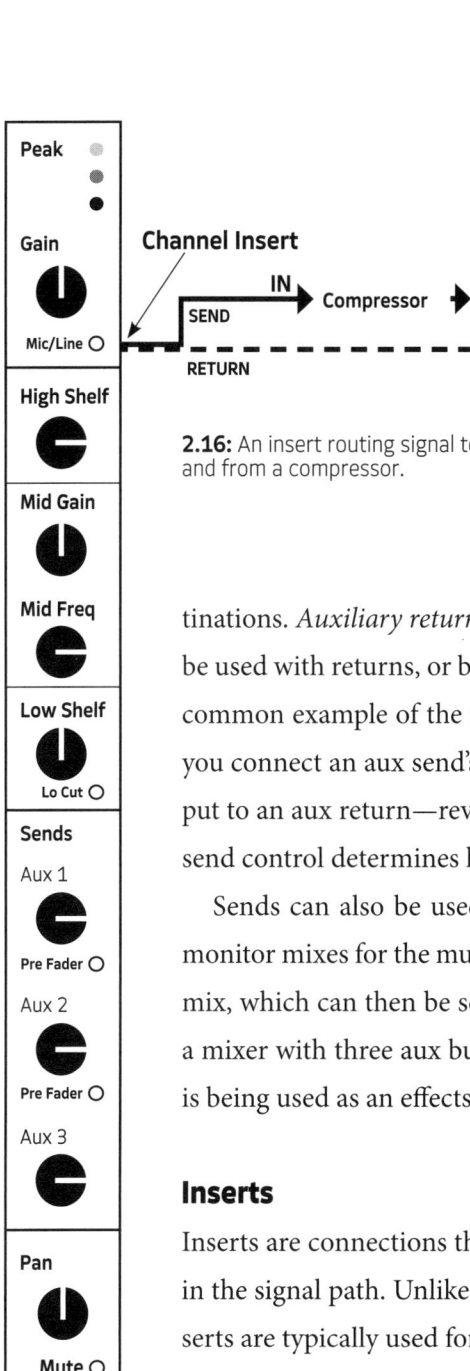

Channel Insert

2.16: An insert routing signal to and from a compressor.

Sends and Retruns

Auxiliary sends let you route a channel's signal to an additional destination without interrupting its flow to the main outputs. Most mixers have identical send controls for every input channel. All the sends with the same name go to the same output. (That is, the aux 1 sends on channels 1, 5, and 9 all route signal to the aux 1 output.) You can connect send outputs to any number of destinations. *Auxiliary returns* are special inputs in a mixer's master section. Sends can be used with returns, or both kinds of connections can be used separately. The most common example of the send-and-return combo is an *effects send.* For example, if you connect an aux send's output to a reverb's input—and connect the reverb's output to an aux return—reverb will be available on every channel. Each channel's aux send control determines how much reverb it will get.

Sends can also be used without returns. This technique is often used to create monitor mixes for the musicians. Each auxiliary bus can be used to create a separate mix, which can then be sent to stage monitors, headphones, etc. Figure 2.15 shows a mixer with three aux buses. Aux 1 and 2 are being used for monitor mixes; aux 3 is being used as an effects send.

Inserts

Inserts are connections that route signal to and from a channel—usually very early in the signal path. Unlike aux sends, they do interrupt the channel's signal flow. Inserts are typically used for adding effects such as compressors, gates, and equalizers (Chapter 5). One "side" (the *insert send*) routes signal to the external device; the other (the *insert return*) accepts signal from the external device. Figure 12.16 shows a typical channel insert connection. You may also find inserts on sub and main mix buses. Unfortunately, inserts are not always available on powered mixers and all-in-ones. With digital mixers, which we'll cover shortly, physical insert connections are often replaced by internal effects that can be "inserted" into a channel.

Direct Outs

Direct outs are connections that take an individual channel's output, independent of either the sub mix or main mix. These do not interrupt the channel's signal to the main mix; instead, they tap into it so that you can route the channel somewhere extra. This is useful for live recording with a multitrack recorder.

The Digital Difference

Thus far, we've focused on analog mixers because they provide a solid foundation for understanding how mixers work in general. For the most part, digital mixers are patterned after analog mixers. They have inputs, gain controls, equalizers, sends, faders, buses, and outputs, just like their analog counterparts. However, there are a few additional considerations when using a digital mixer.

DIGITAL MIXER CONTROLS

Digital mixers have two kinds of controls. Some are *physical* controls that work just like the ones on an analog mixer. A physical control has one job—for example, setting the gain on an analog input—and its setting is "what you see is what you get."

But digital mixers also have data controls (Figure 2.17). They may look like the same knobs and sliders you find on an analog board, but their job is to send a data signal to the mixer's internal computer. Because of this, these controls can do different things at different times. For example, the same fader might control channel 1 or channel 9, depending on how the mixer is set. A trio of knobs may be used for EQ on one channel strip, then switch over to adjust the reverb on another.

2.17: This Yamaha 01v96 is typical of a compact digital mixer. The knobs, buttons, and data wheel shown in the detail above have many jobs, depending on what channel the user chooses to work on. The screen shows what parameters the knobs are controlling at any given moment.

Sharing a "Channel Strip"

Some large digital mixers have full channel strips, with one knob per function, but most compact digital mixers use one set of knobs to control each channel in turn. First, you must select the channel you want to adjust. A display screen shows you which channel is active, and you adjust the knobs accordingly, sometimes using menus to move between different parts of the virtual "channel strip." When you're ready to go on to the next channel, you select it and use the same set of controls.

ANALOG AND DIGITAL SIGNALS

Digital mixers have connections for both analog and digital signals. These are not interchangeable! An analog input can connect to an analog sound source, such as a microphone, instrument, or analog mixing board. Analog outputs can connect to power amps, analog mixing boards, tape decks, the analog inputs on effects devices, monitor speakers, and headphones.

Digital inputs and outputs are only used to connect with other digital gear such as digital mixers, instruments and effects devices with digital I/O (ins/outs), digital recorders, and outboard analog-to-digital converters.

Analog to Digital Conversion

In order for an analog signal to work with a digital mixer, it must be converted into a digital format. This is known as *A/D* (for analog-to-digital conversion). Before the signal can go from the mixer to any piece of analog gear—such as speakers or headphones—it needs to be converted back to analog.

Sample Rate and Bit Depth

The way a digital mixer converts sound to digital data is determined by two important stats: *sample rate* and *bit depth*.

Digital audio systems divide sound into tiny little data segments, called *samples*. The more samples per second, the better the audio quality. A sample rate of 44.1 kHz means that every second of audio is divided up into 44,100 samples. That's the rate used by CDs. Today, it's not uncommon for digital audio devices to have sample rates of 88.2 kHz, 96 kHz and even higher.

Bit depth refers to the number of data bits used to represent a single sample. As with sample rate, higher bit depths usually indicate better sound quality. CDs are 16-bit. As we go to press, 24-bit is the most common format.

SIGNAL ROUTING IN A DIGITAL MIXER

Signal routing in a digital mixer is usually patterned after the signal routing of analog mixers—that is, there are inputs, outputs, buses, sends, returns, etc. The big difference is that a digital mixer may have fewer external connections for individual channels (Figure 2.18). To make up for it, the mixer may have internal effects that can be "inserted" into a channel the way you would insert an outboard signal processor on an analog mixer.

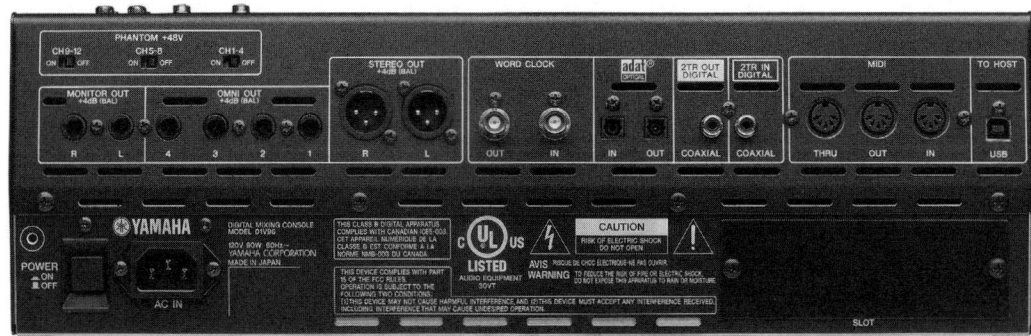

2.18: The rear panel of the 01v96 mixer shown in Figure 2.17. There are a combination of digital, analog, and data connections.

Internal Routing

The internal routing on a digital mixer will usually include sends and returns that patch into onboard effects, as well as internal inserts that can be used to add processors like compression and gating to individual channels. You may also find elaborate solo, mute, group, and submix options at your disposal.

Connecting to Outboard Gear

Every digital audio device runs on a clock that controls its sample rate, and these sample rates need to match for the audio to sound right when two or more devices share a digital connection. (If two digital pieces of gear are connected by analog connections, the sample rate actually doesn't matter; the mixer will treat the other digital device as just another analog source.)

If you do use a digital audio connection between two or more units, make one the clock master and the other(s) the clock slave(s). If two or more devices are left to run on their own internal clocks, they can drift out of sync, even if their sample rate settings match. This can cause distortion, jitter, and other audio problems. Professionals often use a separate clock source, called *word clock*, to control all the digital equipment in a system.

Connecting to a Computer

Many digital mixers can connect directly to a computer via USB or FireWire cables. These connections can be used to send audio between the mixer and computer software. The computer can be added to the mix and used as a virtual instrument, multitrack recorder, effects device, or even a submixer. With the right software, a computer can also be used to store and organize a digital mixer's settings and operate the mixer's controls.

MIDI Connections

The MIDI connections on a digital mixer allow it to connect to other MIDI devices, such as keyboards and sequencers. MIDI cannot be used to send audio, but it can be used to send data. One example: A MIDI keyboard could send a program change to a digital mixer that automatically adjusts its settings. This is useful if your mix is automated in any way.

AUTOMATION

Mixer automation stores the settings of a mixer for later recall. Automation can take two forms: *snapshot automation* and *moving fader automation*.

Snapshot Automation

This kind of automation stores all of a mixer's settings—levels, EQ, sends, returns, etc.—in something called a *scene*. When you recall the scene, every parameter stored in it will be set where you left it. Change scenes, and all the controls change accordingly.

Moving Fader Automation

Moving fader automation changes individual parameters as the music plays. To work, the mixer needs to synchronize to some kind of clock, such as the time code produced by music software or a drum machine. First, you must record a mix by moving the controls. When you play the mix back, the controls duplicate your movements. Such automation is less common in live situations unless you play to prerecorded tracks or synchronize to a click.

COMPUTER-BASED MIXERS

With the right software, a computer connected to an interface with the needed number of physical inputs and outputs can make a formidable mixer. This is the standard in recording studios. Why not live? Well, it can work—but you'll need to make sure your interface doesn't introduce much of a time delay, known as *latency*, when sending sound to and from a computer.

Next Stages

Mixers are standalone units, but you'll find that many pieces of gear have some sort of mixer built in. Electronic keyboards with the ability to layer sounds and add effects; drum machines that combine all of the sounds of the drum kit; keyboard and acoustic guitar amps with separate channels for a mic and instrument inputs; computer software—they all let you control more than one sound at a time and blend them together. And no matter how big or complex a mixer gets, that's really all it's doing.

If the mixer can be seen as the brains of your sound reinforcement system, amplifiers and speakers are the muscle. They're coming next.

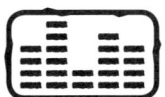

Chapter 3

AMPLIFIERS
AND SPEAKERS

Unless you're playing the most intimate of spaces—like a living room—your onstage sound will almost certainly require amplification. Technically, an *amplifier* is any device that can increase the strength, or *gain,* of a signal. We'll narrow that down slightly by saying that an amplifier (amp for short) is a device that can make a signal loud enough to power a set of speakers.

That's still pretty general. After all, the speakers could be inside a set of headphones (in which case you'd be using a *headphones amp*); they could be the house or monitor speakers of your P.A. system (in which case you'd be using a *power amp*); or they could be in a speaker cabinet used by instrumentalists (in which case you'd be using a guitar, bass, keyboard, or other *instrument amplifier*). And while these amps vary widely in sound, features, and even physical form, they all serve as the bridge between your instrument or mixer and the speakers that bring sound to the audience's ears. Therefore, we'll look at amps and speakers together in this chapter.

Amp Categories

There are a few different ways to categorize amps. You can look at their function (guitar, bass, headphone, etc.). You can look at their physical form (head, combo). You might even look at their circuitry (tube, solid-state) or power rating (low or high wattage). All of these factors are important, but I think function is the best starting point.

In a sound reinforcement system, you can divide amps into two big groups:

1. Amps used for individual instruments.
2. Amps used for the P.A. system.

In most cases, these categories don't overlap. You wouldn't use a power amp designed to drive P.A. speakers for your guitar, would you? No. Well, not usually. As you'll see, there are times when a P.A. component can be part of an individual setup. Even more common, especially for solo performers and small ensembles, are individual amplifiers that can serve as miniature P.A. systems. Before we get into all that stuff, let's take a quick look at how amplifiers work.

How Amplifiers Work

An amplifier uses electricity to boost the signal coming into its input. So, like a mixer, an amp has a signal path. You can divide an amp's signal path into two stages:

1. The *preamp,* which boosts weak signals so that they can be processed by the electronics.
2. The *power section,* which makes the signal strong enough to drive the speakers.

PREAMPS

A preamp—shorthand for *preamplifier*—is the part of an amp that does the most to shape the sound of the microphones and instruments plugged into it. Preamps can either be stand-alone units or they can be part of some other device. In fact, every mixer channel is basically a preamp, which boosts the signal coming from a microphone, instrument, or other source. In that same way, you can look at a mixer as a large and complex preamp.

Preamps usually have controls for gain, as well as some sort of EQ. But they can get far more complex. In most cases, a preamp is used to do more than boost a weak signal; it's supposed to give it some character too.

Figure 3.1 shows a very basic preamp's signal path. The input signal is boosted by a *gain stage* (controlled by the gain control) and fed to a power amp. Before the signal reaches the power amp, a three-band *equalizer* can change the signal's overall tone by adjusting the gain of specific frequencies. Some preamps—especially those built for electric guitar and bass—may have multiple gain stages.

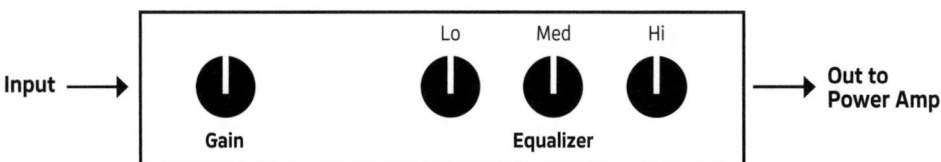

3.1: A basic preamp signal path.

POWER AMPS

A power amplifier takes a signal that's already been boosted by a preamp and gives it the strength it needs to drive a set of loudspeakers. An amp's power is measured in *watts*. In general, the more watts an amp puts out, the louder it can get without sounding distorted. Unlike a preamp, a power amp has few controls. In fact, there's often only one: volume (technically, on most standalone power amps, this is input sensitivity). In most cases, its job is to make the signal louder while doing as little as possible to change its tone.

Obviously, if a preamp can be either a standalone device or a section of a single unit, so can a power amp. As we saw in Chapter 1, standalone power amps are very commonly used in P.A. systems. Figure 3.2 shows a typical power amp used in sound reinforcement. As you can see, there are not a lot of knobs!

3.2: Crown power amp

Standalone power amps like the one shown in Figure 3.2 usually have two channels. You can use the two channels of a power amp to run a stereo mix onstage by feeding the amp the mixer's left and right main outputs. Or you can use them as two mono channels, with one channel taking the mixer's main out and feeding the house system, and another taking an aux or sub out to onstage monitor speakers for the musicians. Figure 3.3 shows the connections on the back of a power amp.

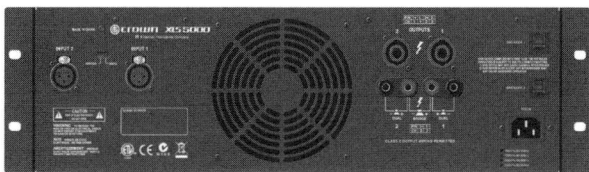

3.3: Power amp connections

On many power amps, the two channels can also be combined into one. This is called *bridging*. Power amps are more powerful when running in bridged mode, but the voltages produced in this mode can be dangerous if the amp is very powerful.

Power Rating and Impedance

When you look at an amplifier's power rating, you'll see more than the number of watts the amp puts out. Another number tells you the *impedance* of the speaker system the amp is driving. For example, if you see "1000 watts per channel @ 4 ohms," that means that each side of the amp puts out 1,000 watts when connected to a 4-ohm speaker cabinet. In this case, the speaker is known as the *load*.

Impedance measures electrical resistance, and it's very important to know about when working with power amps and speakers. To work correctly, power amps *must* be connected to a load that has the proper impedance. In fact, mismatching impedance—or running a power amp without connecting it to a load—can do serious damage to an amplifier. The most important thing to avoid is connecting a load that has too low an impedance. If the amp is rated for 4 ohms, running it into a 2-ohm load can fry its insides. Running it into 8 ohms will cut its power in half—not great, but not as damaging.

Most power amps will also publish a rating for mono/bridged operation. For example, the Peavey CS800 power amp is rated like this:

2040 watts per channel at 4 ohms

1250 watts per channel at 8 ohms

4080 watts at 8 ohms bridged

As you can see, changing the impedance affects the amp's wattage.

COMPONENTS AND COMBOS

If we look at amplifiers and speakers together—and in a practical sense, you really can't have one without the other—we're left with three basic pieces: the preamp, power amp, and speakers. The physical form these three pieces can take varies; they can be three separate

3.4: When a system has a separate amplifier head and speaker cabinet, it's known as a stack. This SWR bass amp is driving two different cabinets.

units plugged together as a chain to form one system, units that combine two pieces (for example, preamp and power amp in one device or power amp and speaker in one device), and units that include all three (such as an all-in-one P.A. system). In sound reinforcement, the mixer feeds one or more power amps, which in turn drive the loudspeakers.

With individual instruments, it's more common for preamp and power amp to be combined into one unit, known as a *head*. Powered mixers are similar. When a head is used to drive a separate speaker cabinet, the two units together are called a *stack* (Figure 3.4), though that term usually applies to guitar and bass amps, not P.A. systems.

Speakers Onboard

Many instrument amplifiers include a speaker in the same housing as the head. These are known as *combos*. Combos usually include between one and four speakers specially designed to work with the head. Figure 3.5 shows a combo amplifier designed for electric guitar.

3.5: A combo amp like this Roland JC-120 has both amp and speakers in one cabinet.

Speaker Basics

Loudspeakers—speakers for short—are the final destination for your signal before it's turned back into audible sound waves. You'll use them to send sound to the audience and you'll use them to hear yourself play and sing. Speakers may be used for the entire group going through your P.A., or for individual instruments coming through your own amps.

Speakers can be used individually or in systems that combine two or more individual speakers, or *drivers*. Drivers are housed in specially built boxes called *enclosures* or *cabinets*, and these cabinets may be combined into one large system, though you have to know a bit about both the amp and speakers to do this yourself. The drivers are often referred to by their diameter, given in inches (i.e., a 12" speaker has a diameter of 12 inches). Figure 3.6 shows a raw speaker driver.

3.6: Speaker driver

TYPES OF DRIVERS

Speaker drivers come in many types and sizes, but for the purposes of onstage sound, you can divide them into three categories: full-range drivers, tweeters, and woofers.

Full-Range Drivers

A *full-range* driver is designed to reproduce low, midrange, and high frequencies on its own. That's a tall order, so most full-range speakers tend to emphasize the midrange at the expense of low and high frequencies. This makes them a bad choice as P.A. speakers—but a great option for electric guitar. Most guitar amps use full-range speakers, such as the one shown in Figure 3.7.

3.7: A full range driver like this Celestion Vintage 30 is commonly found in guitar amp cabinets.

Woofers and Tweeters

Woofers are low-frequency drivers, which handle bass sounds. Woofers used in sound reinforcement generally run from 10" to 15" in diameter. Some instrument amplifiers may have woofers as small as 6" in diameter.

Tweeters are designed to handle high frequencies. While there are some cases where a high-frequency driver might be used on its own, tweeters are almost always combined in a single enclosure with a woofer, as shown in Figure 3.8. A device called a *crossover* splits the signal at a certain frequency, sending the highs to the tweeter and the lows to the woofer.

Subwoofers

Speakers that are designed for ultra-low frequencies are called *subwoofers*. While standard woofers are usually housed in the same enclosure as tweeters, subwoofers are usually in their own, special enclosures—and often have an independent source of power. Subwoofers can get as big as 18 inches.

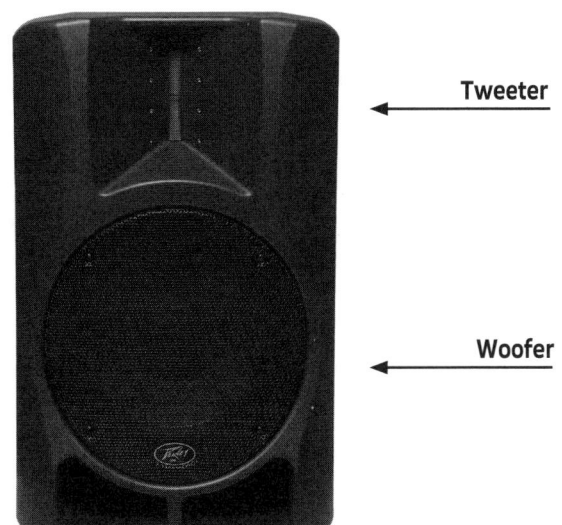

3.8: This Peavey Impulse 12D cabinet houses a woofer and tweeter.

Tweeter

Woofer

Co-Axial Drivers

Co-axial drivers combine a woofer and a tweeter in a single unit, which allows them to produce sound from the same physical location. Figure 3.9 shows an illustration of a co-axial driver.

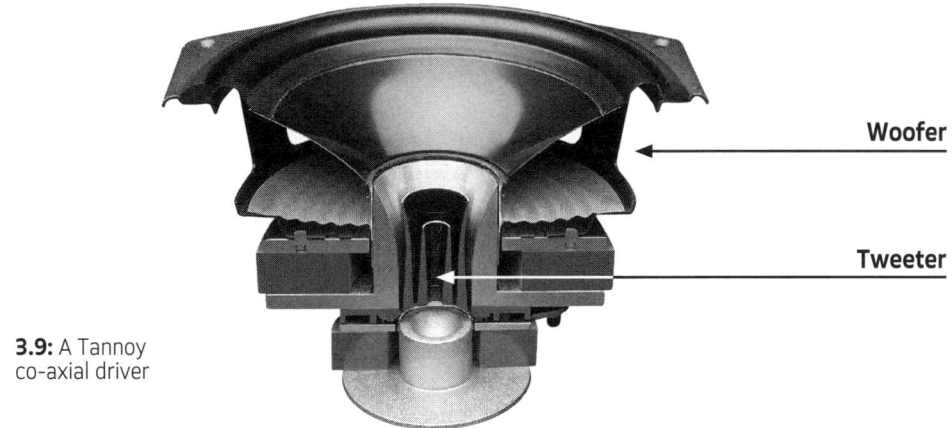

Woofer

Tweeter

3.9: A Tannoy co-axial driver

SPEAKER ENCLOSURES

Speaker enclosures do more than just provide a living space for raw speaker drivers; they actually contribute to the overall sound of the speaker system. That's because the structure of the enclosure interacts with the vibrating speaker and vibrates itself. Depending on how the enclosure is set up, its structure can affect the direction in which the sound projects, as well as the speaker's overall frequency response.

Grilles and Ports

The front of the driver, which is the part that projects sound, is pretty delicate. Therefore, it's usually protected by a barrier known as a *grille*. Some enclosures use grille cloth, a special fabric that lets the sound come through. Others use harder material, such as wire mesh.

Enclosures may be sealed, so that their only openings are for the drivers; or they may

3.10: A front-ported bass cabinet

Driver

Port

have openings called *ports*, which project sound. Ports can help improve a speaker's bass projection, so they're common on P.A. speakers and bass amp cabinets. Figure 3.10 shows a front-ported bass cabinet housing a single 15" speaker.

Guitar amp cabinets can have open or closed backs. An open-backed cabinet isn't the same as a ported cabinet; in fact, it tends to have a little less bass response, but has a wide sound projection both front and back.

Enclosure Configurations

An enclosure can have one driver, or can include a set of them. The drivers can all be of the same type, or may include a combination of different drivers, each designed to handle a specific frequency range.

Speaker cabinets with full-range drivers are described by the number and diameter of the speakers they contain. For example, a 4 x 12" cabinet houses four 12" speakers. Figure 3.11 shows an example.

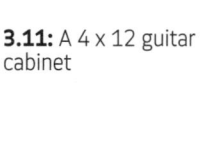

3.11: A 4 x 12 guitar cabinet

Enclosures with woofers and tweeters are described by the size of the woofer and the number of different types of drivers they house. So a two-way 12" enclosure would have one 12" woofer and one tweeter.

ENCLOSURES AND LOAD

One of the most important things to know about any speaker enclosure is its impedance. Remember, the speaker's impedance must match that of the amplifier for the system to work properly. If there's only one driver inside, the impedance is the same as that of the driver. That is: An enclosure housing a single 8-ohm driver will have an impedance of 8 ohms.

If the enclosure has more than one driver, things get more complicated. The enclosure's overall impedance will depend on how the speakers are wired together. Speakers can be wired in *series*, *parallel*, or *series-parallel*.

Wiring Speakers in Series

When you wire two speakers together in series, you must add the impedance of each driver together to get the total impedance of the cabinet (D1 + D2 = Total Load). So if two 4-ohm speakers are wired in series, the cabinet's impedance is 8 ohms.

To wire speakers in series, connect the amp's positive output to the positive terminal of the first driver (Driver A). Driver A's negative terminal connects to the positive terminal of Driver B. Driver B's negative terminal connects to the amp's negative output. This is shown in Figure 3.12. The more speakers you add, the higher the impedance.

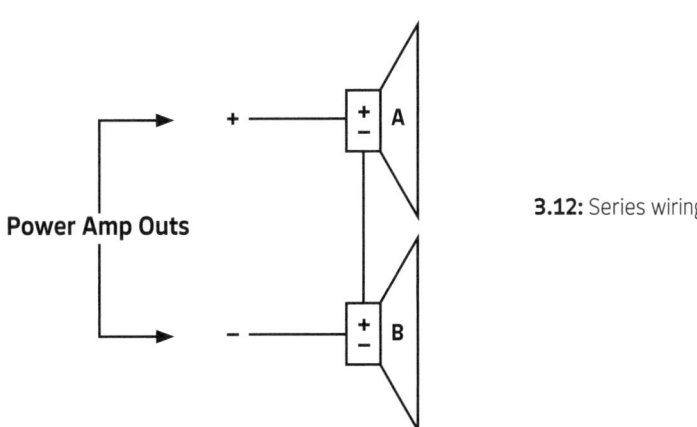

3.12: Series wiring

Wiring Speakers in Parallel

When you wire two speakers together in parallel, the total impedance actually drops. The equation for this is a little more complicated: You have to first multiply the speaker impedances, then add them together, then divide the product by the sum (D1 x D2/D1+D2=Total

Load). It's actually easier than it sounds when you plug in real numbers. Two 8-ohm speakers wired in parallel give you a total of 4 ohms. Here's the calculation:

8 x 8 = 64 (that's the D1 x D2 part)

8+8 = 16 (D1 + D2)

64÷16 = 4

To wire speakers in parallel, you send the positive output from the amp to the positive terminal of one speaker, and the negative output from the amp to that same speaker's negative terminal. You then connect the speaker's positive terminal to the positive terminal of the second speaker. The negative terminals also connect together, and so on. Figure 3.13 shows speakers wired in parallel.

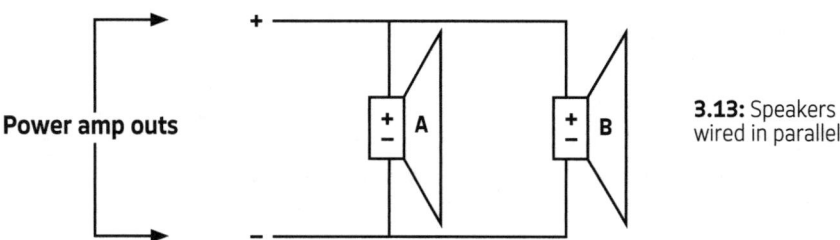

Power amp outs

3.13: Speakers wired in parallel

When there are only two speakers, and they both have the same impedance, you can skip the formula because the impedance will be half as high as the speakers' individual rating. For two 8-ohm speakers in parallel, the total impedance is 4 ohms, because 8 is half of 4.

Series/Parallel Wiring

Series and parallel wiring work great when there are only a couple of speakers, but if you were to wire a larger group of drivers together, the impedances could get kind of crazy. Four 8-ohm speakers in series produce a 32-ohm load. Wire them in parallel, and the load would be 128 ohms. Neither one is very amp-friendly.

The solution is series-parallel wiring. Here, one set of speakers is wired together in parallel, and each speaker in the set connects to another in series, as shown in Figure 3.14.

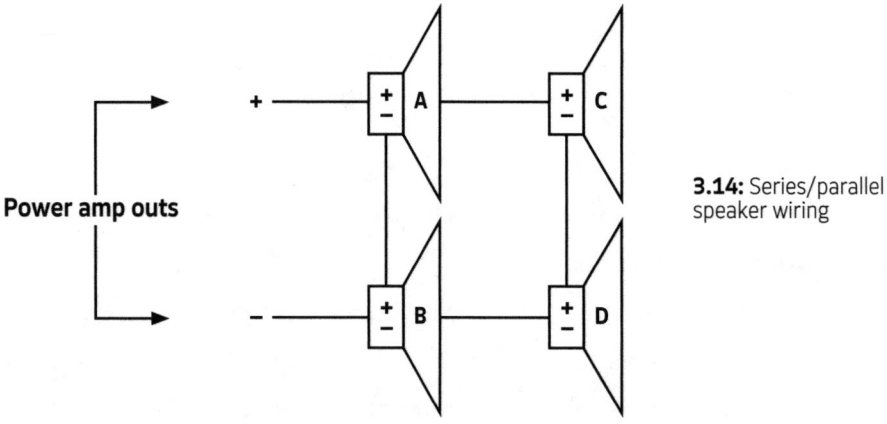

Power amp outs

3.14: Series/parallel speaker wiring

To figure out the total impedance, you need to combine both equations. Let's say we have four 8-ohm drivers (D1, D2, D3, and D4). D1 and D3 are wired together in parallel; D1 and D2 are wired together in series; D3 and D4 are wired together in series.

First, add up one set of speakers wired in series (D1+D2). Next, add up the second set of speakers in series (D3+D4). Plug the total from each of those sums into the formula for parallel wiring. Here's how it would look:

D1 + D2: 8+8=16

D3 + D4: 8+8=16

16 x 16 = 256

16 +16 = 32

256÷32 = 8

Whew!

Powered Monitors

Powered monitors are speaker cabinets that have built-in amplifiers. These are becoming more popular both as house speakers and as stage monitors. Not only do they save you the hassle of carrying around heavy power amps, but they're also designed so that the power and the speakers are perfectly matched. You can connect a mixer directly to a powered monitor, and some models even have built-in microphone preamps and instrument inputs.

Bi-amping

Some powered monitors are *bi-amped*, which means that they have separate amplifiers driving the woofer and the tweeter. This ensures that each driver gets the ideal amount of power.

Wedges

Wedges are speaker enclosures that are used for stage monitors (Figure 3.15). The enclosure's wedge shape allows the speakers to sit on the stage floor and point up at the musicians. Some cabinets can actually be used either as wedges or as conventional upright speakers.

3.15: This JBL wedge has a bracket that allows it to be mounted upright on a stand.

Instrument Amplifiers

Instrument amplifiers play a huge role in onstage sound. Understanding how to choose the right amp for your instrument, set it up for good tone, and then mix it in with other instruments can really enhance your performance.

P. A. systems are generalists: They work with a global range of audio sources. The amplifiers that are designed for instruments are specialists: An electric guitar amp will operate differently from one designed for electric basses, which will in turn sound different from an amp built for acoustic guitars or one made for electronic keyboards.

An instrument amp includes both a preamp section and a power section. Usually these are combined in a single unit, but you can also buy preamps and power amps as separate components. The preamp is a little like the mixer input channels we discussed in Chapter 2. There's an input for the instruments (some also include inputs for microphones), and controls for gain and EQ controls. The sound is shaped in the preamps before it goes on to the master section. Let's take a quick look at amps by instrument, then check out some of the key features you might find on an instrument amp.

ELECTRIC GUITAR AMPS

Electric guitarists have a huge array of choices when it comes to amplifiers. They range in size from combos that will fit in a backpack to stacks that are as tall and wide as an NFL lineman. Some use technology dating back to the earliest days of audio electronics. Others use cutting-edge digital processing.

Most electric guitar amps use full-range speakers that color the sound by reducing extreme high and low frequencies. This makes them a poor choice for instruments other than electric guitar. Figure 3.16 shows a typical electric guitar combo, as well as a stack.

3.16: Peavey combo (left) and Epiphone stack electric guitar amps

BASS AMPS

Like guitar amps, bass amplifiers come in combos or stacks. In order to reproduce the low frequencies of the bass without distorting, bass amps generally produce more output power, in terms of watts, than guitar amps.

Bass amp cabinets may carry one or more full-range speakers, or they may use a combination of drivers for low and high ranges. The cabinets are usually ported to allow the bass frequencies to be heard. Figure 3.17 shows a small bass combo and a bass stack.

3.17: Hartke combo (left) and Acoustic stack bass amps

ACOUSTIC GUITAR AMPS

Acoustic guitar amplifiers are so different from their electric counterparts that they deserve their own category. The most important difference is in the speakers: While electric guitar amps use one or more full-range speakers, which recreate the entire signal from low to high, most acoustic guitar amps use a two-way system with woofers and tweeters.

Acoustic guitar amps often have microphone inputs as well as instrument inputs. These inputs usually have their own individual channel controls, which allows them to be mixed independently.

The microphone input on an acoustic guitar amp can be used for vocals as well as instruments, which allows the amplifier to be used as a miniature P.A. system. Thanks to their ability to reproduce a broad range of frequencies, these amps are also good choices for

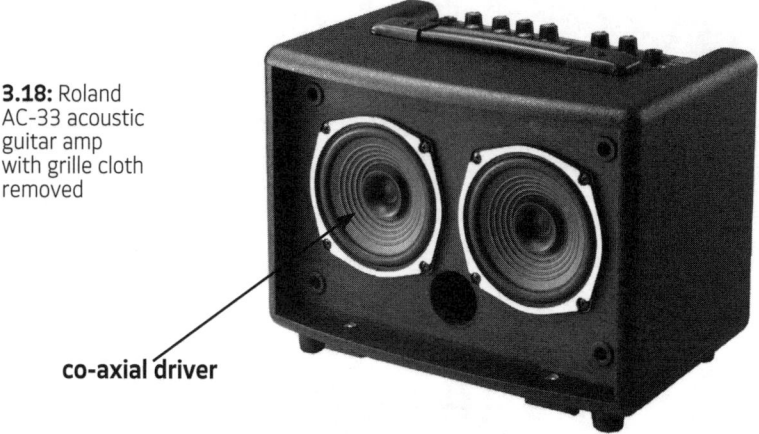

3.18: Roland AC-33 acoustic guitar amp with grille cloth removed

co-axial driver

stringed instruments like mandolin, violin, etc. Figure 3.18 shows a compact acoustic guitar amp. (Note the co-axial drivers on the speakers!)

KEYBOARD AMPS

Like acoustic guitar amps, keyboard amps are basically little P.A. systems. They usually feature several independent channels, and may have stereo inputs as well as one or two mic inputs. They tend to be larger and produce more bass response than acoustic guitar amps. Figure 3.19 shows a four-channel amp designed for keyboards.

3.19: The Peavey KB5 combo amp for keyboards has four independent channels. The detail above shows its control panel: Each channel has its own volume and EQ knobs.

Instrument Amplifier Features

Even within a single one of these basic instrument amp categories, you'll find a huge range of features. Some amps are designed to deliver a very identifiable sound. Others are designed to let the instrument sound as pure and uncolored as possible. Various amps for the same instrument may use different types of technology or offer very different kinds of controls. Let's look at some of the key features that give amps their personality.

CIRCUITRY

The type of circuitry inside an amplifier has a huge influence on its sound. You'll find amps that use vacuum tubes, amps that use transistors, and other amps that use a combination of both. Some use digital technology to emulate the sound of tube and solid-state circuits.

Tube Amplifiers

Once, all amps used vacuum tubes in their circuits, but solid-state circuits have largely replaced tubes today. Yet some musicians still love this old-school technology because tubes impart a unique character to an amp's sound. Different types of tubes have different sound qualities. Some, like the 12AX7, are known for producing a lot of gain, and are therefore often used in the preamp section of guitar and bass amps. Others, like the 6L6 and EL84, are used in the power section of various circuits.

Electric guitar amps are the most popular category for tubes, but tubes are also found in some bass amps, acoustic guitar amps, and even some power amplifiers. You'll also find tube preamps for instruments and microphones. These can be fed to mixer channels or power amps. Tube amps can sound especially good when their gain gets high enough to saturate or distort. Many tube preamps are designed for just this purpose.

Solid-State Amplifiers

Solid-state amplifiers use transistors in their circuits instead of tubes. They're lighter and less expensive to maintain than tube amps—one reason why solid-state models are often a player's first amp. Some solid-state circuits, like the famous Roland JC-120, are designed to sound clean and not imitate the gain characteristics of the 12AX7 tube. The ability to get loud without distorting is known as *headroom*. The headroom offered by solid-state amps makes them good choices for bass, acoustic instruments, keyboards, and power amplification.

However, some solid-state amplifiers, especially those designed for electric guitar, emulate tube amps. And while some may consider solid-state circuits to be inferior to tubes, it's worth noting that most stompbox effects are solid-state. Therefore, even those who use tube amps also have solid-state gear in their signal chains. Some amps and preamps, called hybrids, combine tube and solid-state circuits.

Digital Modeling Amplifiers

Digital modeling amplifiers combine a digital preamplifier with a solid-state power section. The digital preamp emulates the sound of a range of "real" amplifiers. These devices can produce a huge array of sounds and usually include built-in effects. Digital modeling amps can be found as standalone devices or in software that runs on a computer, such as Peavey's ReValver, which actually allows users to assemble their own virtual amps out of tubes and circuits emulated in the software.

PREAMP

Most amps have two level controls: one for the input gain, the other for the output or *master volume*. The input gain control is found in the preamp section, while the master volume control is in the power amp section.

Gain

On a guitar or bass amp, the input gain can determine how clean or overdriven the amp sounds. Low gain produces a clean sound. High gain produces more saturation; this is known as *overdriving* the amp. The idea is that you can set the amp to have the amount of overdrive you want and then control the loudness with the master volume. With sources like acoustic guitar or keyboard, you'll probably want to set the gain low.

In either case, the gain control should be adjusted to match the level coming from the source. If an instrument has weak pickups, you might need to crank the gain to get even mild overdrive. If it has powerful pickups, you might have to back off the gain to get any semblance of a clean sound. The amp may have a high/low gain switch or use specific inputs to match sources with strong or weak signals.

Channels

On some amplifiers, the preamp section offers more than one channel. A *channel-switching* amp, like the one shown in Figure 3.20, lets you use each channel to create a different sound

3.20: A Blackstar channel-switching tube amplifier head

and toggle between them using a footswitch. This is a popular design for electric guitars because it lets you set up separate tones for rhythm and lead parts. Typically, one of the channels will produce a clean sound, while the other is usually a *high-gain* channel designed for distorted sounds. Some amps have a third channel, sometimes called the "crunch channel," for a moderately overdriven sound.

On a multi-channel acoustic guitar or keyboard amp, all the channels can be heard at the same time. Each has its own volume control, allowing you to balance the channels to create a final mix.

Tone Controls

The bass, midrange, and treble controls help shape the tone. On some amps, especially older tube designs, these controls also affect the gain and overdrive. (In other words, turning the tone controls to zero is like turning the volume all the way down.) Some amps also have switches for brightness (which adds treble) and depth (which adds bass or low midrange), as well as switches that change the way the midrange control works. You may also find a *presence* knob offering additional treble control.

The tone controls can be *passive*, which means they only cut signal, or *active*, which means they both boost and cut signal.

Onboard Effects

Amplifiers have had built-in effects for decades. Tremolo (sometimes misnamed "vibrato") and reverb were an important part of many vintage amps, especially those built by Fender, Ampeg, and Vox. Roland's JC-120 offered a stereo power amp and a built-in chorus circuit to enhance the stereo effect.

Lately, amp builders have started adding digital effects as well. These can include the traditional effects mentioned above, as well as delay and other modulation effects like flanger, rotating speaker emulation, and more.

Effects Loops and Preamp Outputs

An *effects loop* is similar to the inserts and auxiliary sends and returns found on mixers. These connections allow you to route a signal from the preamp to an effects device and then back to the amp.

A preamp output lets you plug your amp into a P.A. system—or another power amp—so that it can be further amplified without need of a mic. On guitar amps, these preamp outs often include a built-in circuit that imitates the sound of the speaker cabinet. Without such speaker emulation, a guitar amp's preamp outs can sound very harsh. On acoustic guitar, bass, and keyboard amps, the preamp out may be a balanced line designed to run to a P.A. mixer.

POWER AMP

The power section of an amplifier boosts the preamp's signal and feeds it to speakers. For tube guitar and bass amps, the power section can interact with the preamp to influence the overall tone. For most solid-state amps, the power section is designed to boost the signal without adding much color.

Amplifier power is measured in watts, but the raw numbers have to be taken in context. A 50-watt tube guitar amp is more than loud enough to fill a small auditorium. In fact, you may find that it's too loud to turn up past "2." On the other hand, a 100-watt bass amp isn't insanely powerful. The extra wattage is needed to produce the bass frequencies cleanly.

USING A P.A. AS AN INSTRUMENT AMP

In live sound, instruments like bass, acoustic guitar, and keyboards are often fed directly to the P.A. system using a device called a *direct box* or D.I. Therefore, a small all-in-one P.A. like the one in Figure 3.21 can actually be used as an individual amplifier. Many of these have inputs tailored for instruments, so you don't even need to use a direct box. Their controls may be limited, however, so you might use a separate preamp to shape the sound. This is especially true if you're using the P.A. for electric guitar, which usually needs the influence of the amp to produce good tone.

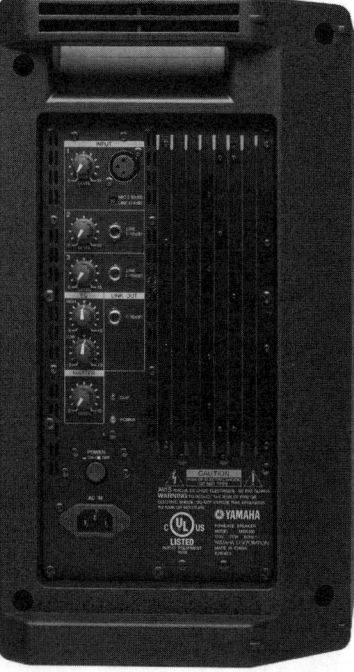

3.21: A small all-in-one powered P.A. can be used as a personal instrument amplifier.

Choosing an Amp

Now that you know a little bit about features, what kind of amp do you need? That's a complicated question and it has many answers.

Some musicians gravitate towards one type of amp based on the style of music they play. A heavy metal guitarist might want a high-gain stack amp. A funk bassist who uses the slap/popping technique might want a rig with plenty of watts driving both woofers and tweeters. A violinist might choose a solid-state acoustic guitar amp with very accurate top end. A blues harmonica player might choose a small tube amp that overdrives easily at low volume. A keyboardist might need an amp that has lots of channels but few other features, opting instead to shape the tone using the keyboard's built-in controls. A singer/songwriter might decide that an all-in-one P.A. might be better than having an amp, since it can handle lots of different instruments as well as vocals.

As your sound becomes more clearly defined, you may find that a specific amplifier is right for you. You may even combine several into a complex rig. But for now, you should know that you can get a pretty good sound out of almost any amp (assuming it's in working order) as long as you set its controls correctly. In fact, you may get better results from an inexpensive amp than from a bigger, more expensive model that's designed for high volume. Some of these can sound really wimpy when you turn them down!

Next Stages

The best way to learn about amps is to try out as many as you can. Experiment with the controls and get to know what sounds good to you. It's important to remember that all amps and preamps are suited to different sounds and sources. We'll look at one of the key ways that sources find their way into an amplifier's signal path in Chapter 4, where we discuss microphones and pickups.

MICROPHONES AND PICKUPS

Microphones and pickups: They seem like the simplest parts of a sound reinforcement system. After all, what's there to know? You sing, speak, or play into one end, plug a wire into some "real" equipment—the stuff with the knobs and sliders—and sound comes out of a speaker.

But mics and pickups are important because they're the first step in the transformation of the music you make into the the sound the audience hears. Professional sound engineers will tell you that choosing and setting up microphones is among their most important jobs. Experienced players will tell you that the pickups on an instrument can completely transform its sound.

The truth is, not every type of microphone or pickup is right for every job. Onstage, the wrong mic choice and placement can cause howling feedback that will make your audience want to run out of the room. A mismatched pickup can destroy the sound of even the best instrument. The more you know about various types of microphones and pickups, the better you'll be able to choose the right tool for the job at hand.

Types of Microphones

Microphones come in many shapes and sizes. There's even one built into the laptop I'm using to write this. But most can be divided into two categories: *dynamic* and *condenser*. Yes, there are ribbon mics (technically, these are a type of dynamic mic) and unusual things like boundary mics. But the majority of the microphones you'll encounter will fall within the two main families we're about to explore.

DYNAMIC MICROPHONES

Dynamic mics are the most common for live performance because they're rugged and relatively simple in construction. Dynamic mics require no external power: You can simply plug one into a mixing board and it will work pretty well.

A dynamic mic works when air pressure moves its *voice coil* (the element inside of the mic that sits inside a protective screen). The coil's reaction to the sound produces electrical energy. It's similar to the way a speaker works, but in reverse.

Generally, dynamic mics are less sensitive than condenser mics. They're known for being able to handle loud sound pressure levels (SPLs), which is one reason why they're used so often on drums, electric guitars, and screaming singers.

Dynamic mics come in many varieties. Some are designed for a specific instrument (for example, several manufacturers make drum packs that include a mic for each drum in a set); others are more general.

Two of the most popular are the Shure SM57 (Figure 4.1), which is probably the most common guitar amp mic in the world, and the Shure SM58—similar to the 57, but with a built-in pop filter for vocal use.

4.1: Shure SM57 and SM58

These two mics alone can get you through almost any gig. Actually, you can use an SM58 on instruments as well. Shure also makes a mic that's very popular for harmonica called the Green Bullet (Figure 4.2); it's designed to plug into a small amplifier instead of a mixer.

4.2: Shure
Green Bullet

Other popular dynamics include the Sennheiser MD421 (especially for horns and drums). Sennheiser's e600 series is also popular for drum miking (Figure 4.3).

4.3: Sennheiser
e604 drum mic

The Electro-Voice RE-20 is often used on kick drums. Peavey, Audio-Technica, Audix, and Blue all make well-regarded dynamic mics; Royer's ribbon mics are also well-liked.

There's no rule that says you can't use a dynamic mic on quiet instruments such as acoustic guitar and violin, but they're not always the ideal choice. Used correctly, condenser mics can work better in such situations.

CONDENSER MICROPHONES

Condenser mics are more sensitive than dynamic mics. That doesn't mean that they'll break if you play punk rock into them. But it does mean that they're better suited to certain kinds of material and environments. For example, condenser mics are commonly used in the studio to record vocals. "Studio" condensers are so sensitive that they may be prone to feedback onstage unless they're handled carefully. However, there are some condenser mics specifically designed for onstage vocal performance, such as the Shure Beta7A and the Neumann KSM-105; these have tight pickup patterns and additional internal shock absorbers that help prevent unwanted noises from coming through the mic's housing.

Condenser mics are more commonly used for instruments in a live situation—especially acoustic instruments (guitar, violin, piano, etc.), drum overheads, and hand percussion—either mounted on stands or as clip-on mics that attach directly to the instrument's body. Clip-ons are particularly popular with wind and string players. You can also use one or two condensers to capture an overall stage sound—useful for a choir or string ensemble, for example. If you want to record your performance, placing a condenser or two in the room can work very well, especially when you blend this with a direct feed from the mixer.

Condenser mics have a charged element, called a *diaphragm*, which picks up sound waves in the air. A condenser mic requires power in order to run. This is provided either by a battery (common on lower-cost condenser mics) or *phantom power*, which is supplied by the mixer via the mic cable. With phantom power, you need only plug the mic into the mixer and activate the power. No other device is necessary. (In fact, if a condenser mic doesn't seem to be working, the first thing to do is check phantom power at the mixer. Just make sure the speakers are muted because turning phantom on can produce a loud pop.)

Condenser mics come in two basic varieties: small-diaphragm and large-diaphragm. The properties of the diaphragm have a lot to do with how a condenser mic captures sounds and will often determine its application.

SMALL-DIAPHRAGM CONDENSERS

Small-diaphragm condensers (Figure 4.4) have relatively small elements, which makes them quick to respond to air pressure. Consequently, they're good at capturing the attack of percussion or the nuances of an acoustic guitar. They also work well on drum overheads and as recording mics, especially when used in matching pairs (a set of two mics paired off at

4.4: AKG C1000 small-diaphragm condenser

the factory because they sound almost identical). Popular examples include the AKG C1000 (previous page), the Neumann KM184 (Figure 4.5) and the Shure KSM137, though there are many other models to choose from. Such mics are sometimes used in matched pairs.

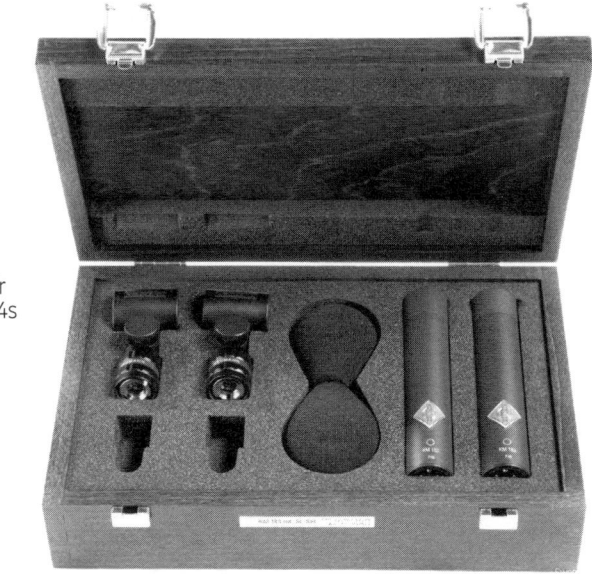

4.5: A matched pair of Neumann KM184s

LARGE-DIAPHRAGM CONDENSERS

Large-diaphragm condensers have a bigger element than their small-diaphragm cousins, and generally offer a more full-bodied sound. These mics are the workhorses of professional studios, but they're less common onstage. However, you will occasionally see them on acoustic instruments, as "ceiling mics" capturing a complete ensemble, and as drum overheads.

Some large-diaphragm condensers, such as the AKG C414 (Figure 4.6) and Neumann U87, feature switchable polar patterns, which means they can be used in a number of different ways. We'll discuss why that's important a little later in this chapter. The best thing about a large-diaphragm condenser (or at least a good one) is that it tends to have a wide response curve and can capture low and high sounds very well if you position it right.

4.6: AKG C414

Mic Characteristics

Sing or play into a handful of different mics, and you'll probably notice that they all sound different. The sound going into the mic doesn't change, so why does the audio coming from the speakers sound brighter with one mic, louder with another, bassier with a third, and so on? This is due to a lot of factors.

Mics are like instruments and amps in that various models have their own sonic personalities. Microphone manufacturers publish data about their products to give you an idea of how they'll sound. And while this information won't tell you subjectively if you'll like dynamic mic A better than dynamic mic B, the following three factors offer a good starting point for understanding how every mic behaves: *pickup pattern* (also known as *polar pattern*), *sensitivity*, and *frequency response*.

PICKUP PATTERN

The pickup pattern tells you how a mic's element responds to sounds coming from various directions. Some mics are focused in one direction, and they reject most of the sound outside this focus area. Others can capture sound from any angle around the element. Directional mics are best for live performance because they isolate the source (for example, the singer's voice) from competing sounds (for example, the guitar amp to the singer's right). Most stage mics have a single pickup pattern, but some can be switched for different degrees of directionality. Pickup patterns fall into four categories: *cardioid*, *super (or hyper) cardioid*, *omnidirectional*, and *figure 8*.

Cardioid Mics

The cardioid pattern is the most common. It's very directional. A cardioid mic will pick up the sound most strongly from right in front of the element, with a little bit coming from the sides. Cardioid mics are used for everything from vocals to instruments to amps.

Figure 4.7 shows a cardioid pattern. The dark line shows that this mic picks up sound most strongly from the front (the top half of the circle), with some coming in from the sides and almost none from the back (the bottom half of the circle).

4.7: A cardioid pickup pattern captures sound in front and to the sides of the mic's capsule.

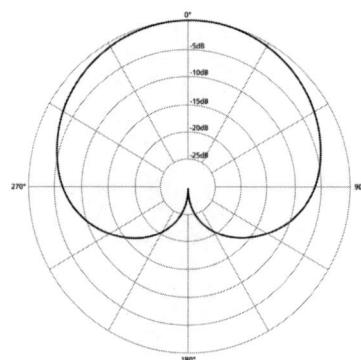

Hypercardioid and Supercardioid

The supercardioid (also known as hypercardioid) pattern (Figure 4.8) is even more intensely focused on sound coming from the front than a cardioid mic. This pattern lets in a very small amount of sound from directly behind the element. Condenser vocal mics designed for stage use tend to have supercardioid pickup patterns.

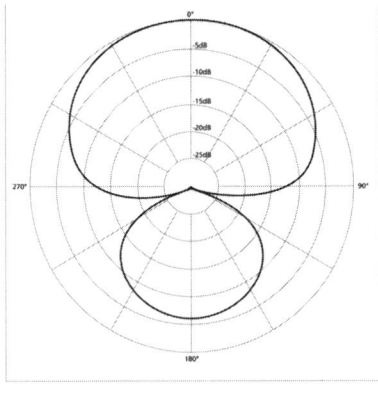 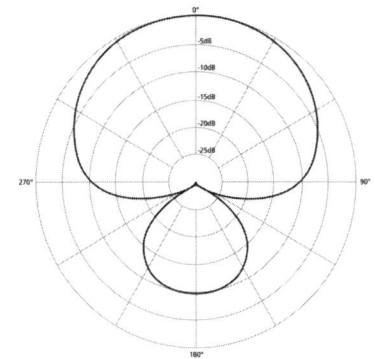

4.8: Supercardioid (left) and hypercardioid patterns are similar to cardioid, but they capture slightly more sound from behind the mic, and less from the sides.

Omnidirectional

An omnidirectional pickup pattern captures sound from all around the element. Unlike cardioid, an omni mic has no direction; it has *all* directions (Figure 4.9).

Omni mics aren't much use onstage, but they do make good room mics for capturing live performance. Many handheld recorders have omni mics built in (or offer the option of switching the mics into omni mode).

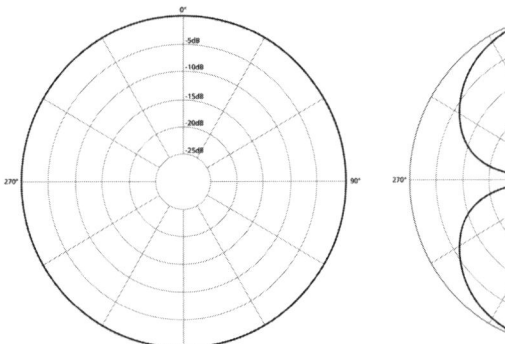 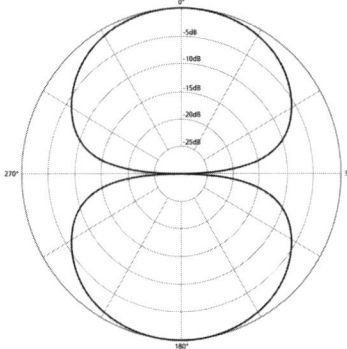

4.9: The omni pattern captures sound in all directions.

4.10: The figure 8 pattern captures signal in front and in back of the capsule.

Figure 8

The figure 8 pattern (Figure 4.10) is a little like having two cardioid mics back to back. It captures sound directly in front of the element and directly behind it. Note that a figure 8 pattern is not the same as a stereo mic; if you record two people through a figure 8 mono mic, their voices will be blended on one channel. These aren't used that often onstage.

SENSITIVITY AND SOUND PRESSURE LEVEL (SPL)

A mic's sensitivity refers to its ability to capture very quiet sounds. Is a more sensitive mic better? Only for some applications, such as quiet instruments or instruments with a wide dynamic range. Onstage, a mic that's too sensitive can be a liability. It can distort when it's confronted with loud sounds, feed back, and be hard to isolate.

As we mentioned earlier, condenser mics tend to be more sensitive than dynamic mics. Therefore, dynamics are generally better suited to handling loud sources that are really close to their capsules. Condensers are almost always a better choice for quieter material or distant sound sources. That said, when used correctly, a sensitive condenser may be able to handle the high SPLs produced by loud sounds like drums and amplifiers. Some condensers mics are equipped with switches called *pads,* which reduce the mic's output a little bit and help it to reproduce loud sounds without overloading.

FREQUENCY RESPONSE CURVE

When you get a new microphone, check to see if there's a strip of paper in the package showing some squiggly lines drawn on a graph. That line shows the mic's *response curve*: how that mic captures different frequencies. Most mics tend to be about even along most of the curve, with some particular ranges being more strongly emphasized. For example, a mic that has an upper midrange "bump" (sound engineer jargon for an increase in level) can work really well for vocals and guitar amp. A mic that has a drop in the lower frequencies won't be as good at capturing bass—but it may help prevent rumble from the stage coming through the mic stand. Figure 4.11 shows the response curve of a dynamic vocal mic. Figure 4.12 shows the response curve of a mic intended for kick drum and bass. As you can see, the bass drum mic has a much stronger response in the low frequencies. You'll also notice that there are several different lines with measurements next to them.

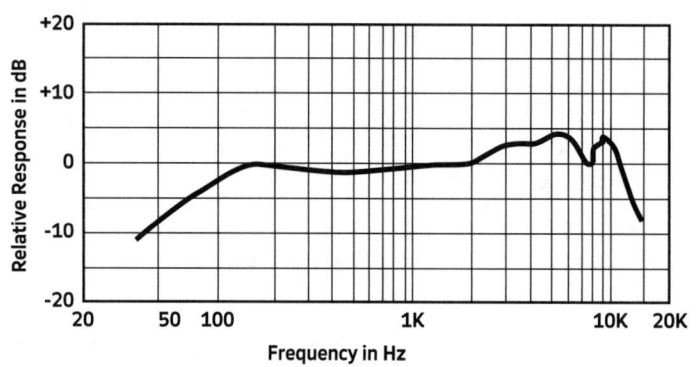

4.11: A vocal mic's response curve

4.12: A bass drum mic's response curve

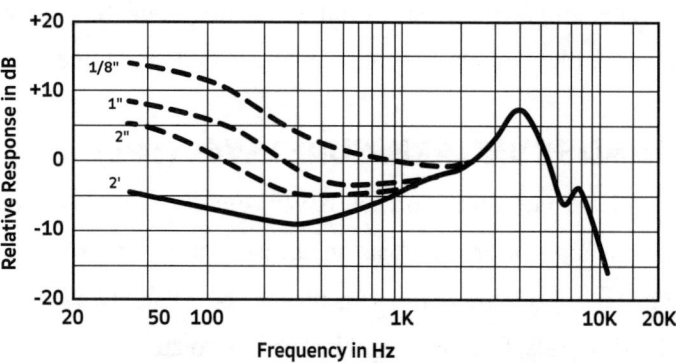

Proximity Effect

A mic's response curve can change depending on how close the source is to the element. When a source is very close to the mic, it can create what's known as *proximity effect*. This refers to how the mic's response changes as the sound gets closer. Most mics have some sort of proximity effect, but it's more pronounced on some. Usually, the effect is an increase in low and low-midrange response. When a singer "eats" the mic, he or she changes the way the mic sounds—which can help some voices sound deeper and others sound dull. You can use proximity effect to bring out more thump in a tom-tom or guitar amp, or to get more warmth out of an upright bass. Figure 4.13 shows a close-up of the SM58's response curve. Note how the lower midrange is boosted when the source is closer.

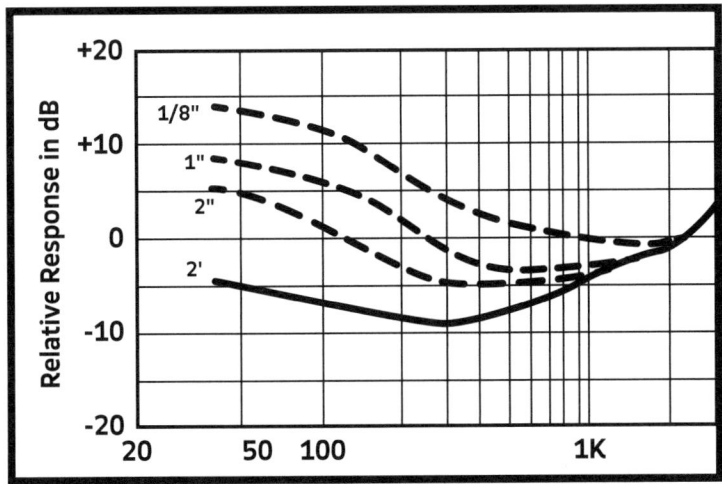

4.13: Here's a close-up view of the lower frequencies shown in Figure 4.12. Notice how the mic's response curve changes: As the source moves closer to the mic, bass frequencies come through at a higher level. This is known as the proximity effect.

Proximity effect can sound good on some sources but lousy on others. One of the challenges of miking an acoustic guitar, for example, is that a mic can sound boomy if it's too close to the instrument.

ADDITIONAL FEATURES

Most mics are very simple and have no controls on them. The mixer is used to set the mic's loudness, shape its tone, etc. However, some mics have switches that change the way they operate.

On/off switch: This is becoming more common on professional mics, and is especially useful for performers who are working without a sound engineer because it lets you mute the mic between sets, during instrument changes, and so on.

Pad: We mentioned this earlier. A pad is a switch that reduces the mic's output to prevent it from overloading the mixer when confronted with very loud source signals. This is more common on condenser mics.

Switchable pattern: This changes the pickup pattern of the mic—for example, turning a cardioid mic into an omni. This has limited use in live applications.

Bass roll-off: This changes the mic's response curve by reducing the amount of low-frequency information it picks up. This can be very useful when the mic is being plagued by stage rumble.

Mounting a Mic

Mics can be mounted several ways, and while it might seem pretty obvious that you snap the mic into a clip and stick it on a stand, there are actually a number of things to consider.

Handheld

Just about any mic can be held in your hands. But that doesn't mean every mic can be considered a *handheld* mic. Vocal mics are designed to mount on stands and be handheld. Therefore, they have extra shock absorbers to reduce *handling noise*—the unwanted sound that transfers from the hands to the mic's element. Figure 4.14 shows the inside of a hand-held vocal mic.

Mic Clips and Shock Mounts

Most mics come with a clip that's designed to hold it securely. The clip screws to the end of the mic stand. (You should make sure that the diameter of the clip fits the stand you're using; not all do.)

Because mics come in different shapes and thicknesses, a clip that fits one model may not work with another. Mic clips tend to get lost and broken, so it's good to have some spares. You might consider a couple of universal clips like the one shown in Figure 4.15.

In addition to basic clips, some mics come with *shock mounts*—devices that isolate the mic from vibrations coming through the stand. These are most common in recording studios, but can be handy onstage as well. Figure 4.16 shows a shock mount.

Headset Mics

Headset mics like the one shown in Figure 4.17 are compact enough to be worn by the player. They're especially handy for musicians who find a mic stand confining, such as drummers, keyboardists who might bop around among a collection of instruments arranged in an "L," and other players on the move.

Mic Stands

Mic stands come in straight and boom types. Boom stands are good for singers who play instruments because they let you position the mic close to your mouth without the stand getting in the way of your playing. Boom stands are also useful for drum overheads, and for positioning mics on sources that are low to the ground, such as kick drums and guitar amps. Booms can be added to straight mic stands when necessary.

4.14: Interior view of an Audio-Technica handheld mic showing shock absorption

4.15: Universal clip

4.16: A Neumann U87 in a shock mount

4.17: Headset mic

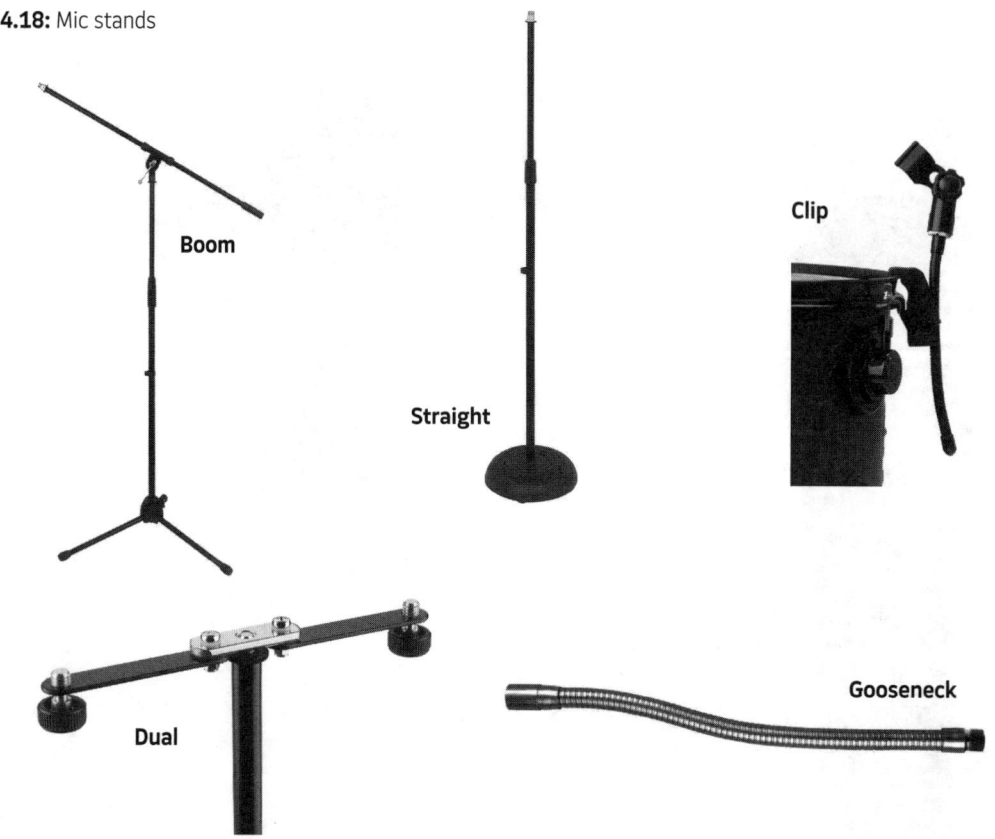

4.18: Mic stands

Boom

Straight

Clip

Dual

Gooseneck

Straight stands are less expensive than boom stands, are easier to pack, and can work fine for many applications. Shorter straight stands are good for low-to-the-ground sources like kick drums or guitar amps.

You can make a straight stand more versatile by adding a flexible metal tube called a *gooseneck*. This can be positioned at an angle away from the stand. Figure 4.18 shows a range of mic stands designed for stage use.

Clip-On Microphones and Other Mounts

Instead of being clipped to independent stands, mics can also be mounted directly on a source. Some mics are designed expressly for this purpose. Figure 4.19 shows a clip mic designed for violin. You'll find similar models for viola, cello, upright bass, and guitar.

4.19: A DPA clip-on violin mic with mounting hardware

Other Accessories

Most mics come with padded bags, but if you plan to gig a lot, you should consider a case like the one shown in Figure 4.20.

Wind screens are also popular. These foam coverings are designed to help tame wind noise—useful for singers and some wind players. Figure 4.21 shows an example. These are not the same as the external pop filters you see on studio mics (Figure 4.22). Pop filters help eliminate the popping "p" sounds known as 'plosions, but they're not all that common for stage use.

4.20: This waterproof SKB case can hold six microphones of various models.

4.21: Shure wind screen

4.22: A pop filter

Pickups

Microphones do an excellent job of capturing sound, but they're not always practical. They can restrict movement. They're also prone to feedback. A *pickup* is a device that connects directly to an instrument, reads its vibrations, and converts that sound energy into an audio signal. Guitar pickups first appeared in the 1920s and eventually led to the creation of the solidbody electric guitar and electric bass, the first popular stringed instruments to rely more on electronics than body resonance for their tone. Later, pickups were developed to capture body resonance as well, and many of these are popular for acoustic instruments.

MAGNETIC PICKUPS

Magnetic pickups are transducers that use magnets (known as *pole pieces*) wrapped in a coil of wire. When a metal string vibrates over a magnetic pole piece, the magnet and wire coil react to create a small electrical voltage. This signal passes through the instrument's electronics before feeding an amplifier's input.

Magnetic pickups can only work if the strings have metal in them. Models designed for electric guitar work best with steel and nickel strings. Acoustic guitar models do a good job of capturing the sound of the bronze, brass, and phosphorus strings those instruments use. You can't use magnetic pickups with nylon and gut-stringed instruments—the magnets don't react strongly enough to their vibrations.

Single-Coil Pickups

The earliest magnetic pickups used a single coil of wire. This design produces good sound—single-coils are especially prized for their clarity in the high and low registers. Unfortunately, single-coil pickups also produce a hum at 60 Hz (known as 60-cycle hum). Single-coil pickups like the Stratocaster-style pickups shown in Figure 4.23 remain popular today, but they can get noisy onstage, especially if you're plugged into an electrical circuit that has a lot of interference on it (lights and dimmers can be especially problematic).

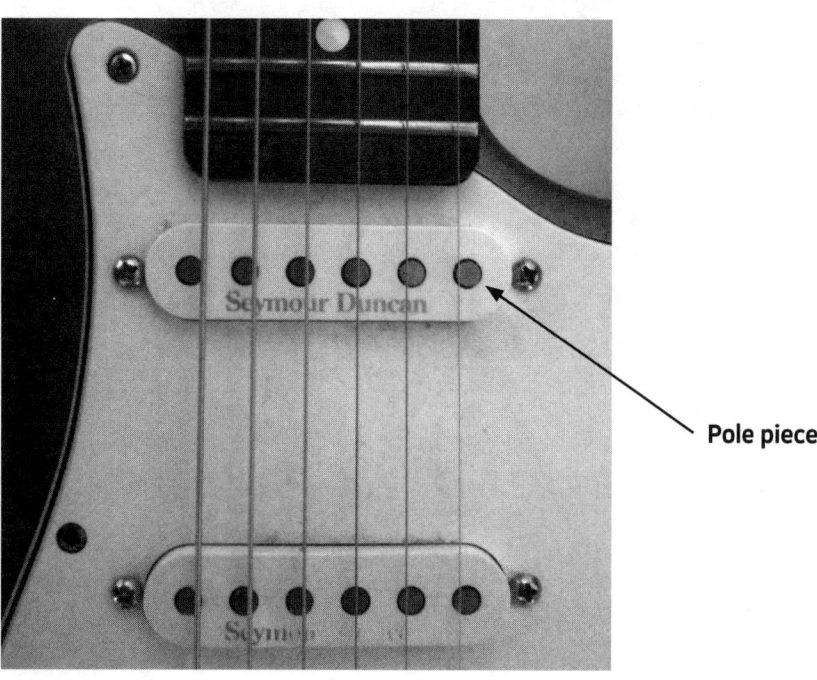

4.23: A guitar equipped with two Strat-style single-coil pickups

Pole piece

Humbucking Pickups

In the late 1950s, a designer at Gibson Guitars named Seth Lover came up with a way to get rid of 60-cycle hum. By combining two evenly matched coils and reversing their polarity, he eliminated the hum and produced a very powerful sound.

Gibson's humbucking pickups (Figure 4.24) became popular in the 1960s, when rock musicians started to use loud amps with lots of distortion, and today, many manufacturers and guitar designs use dual-coil pickups based on those early humbuckers. They're known for having more midrange and a higher output than single-coil pickups, though there's really so much variety among pickups today that it's hard to generalize.

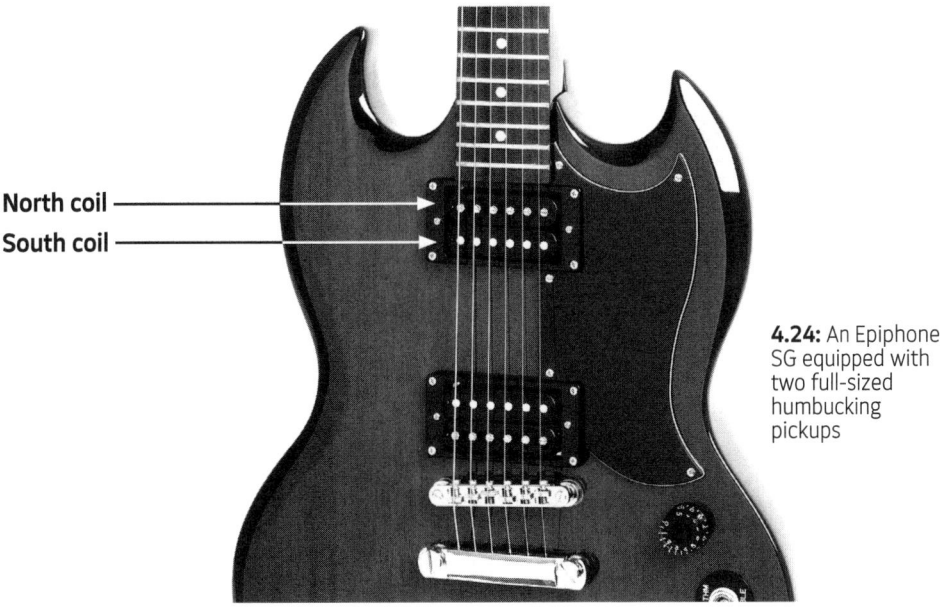

North coil
South coil

4.24: An Epiphone SG equipped with two full-sized humbucking pickups

Hum-Free Single-Coils

Some players still value the single-coil tone, but dislike the noise those pickups produce. This has led to the development of pickups that look and and sound like single-coils, but which also cancel hum. You'll also find noise-canceling circuits that work with existing single-coil pickups; they basically fool the pickup into thinking it's a humbucker.

Active Pickups

Active pickups use internal circuitry to boost the pickup's output. These pickups offer low noise and high gain. The active circuits—which require a battery to run—can also allow the player to access a lot of different tones. Active pickups can be found on both guitars

Pickups and Amps

Guitar and bass amplifiers are specifically designed to work with magnetic pickups, and vice versa. You may notice that the tone is a little different if you plug a guitar directly into an amp's input, versus plugging it into a preamp first. Amps also react a little differently to passive magnetic pickups vs. active pickups. Some players like the way an active pickup pushes an amp, but you might find that you lose some headroom from the amp when running active pickups. Try switching to an input designed to handle the higher output. If the amp has a passive/active switch, try both positions to see which is best.

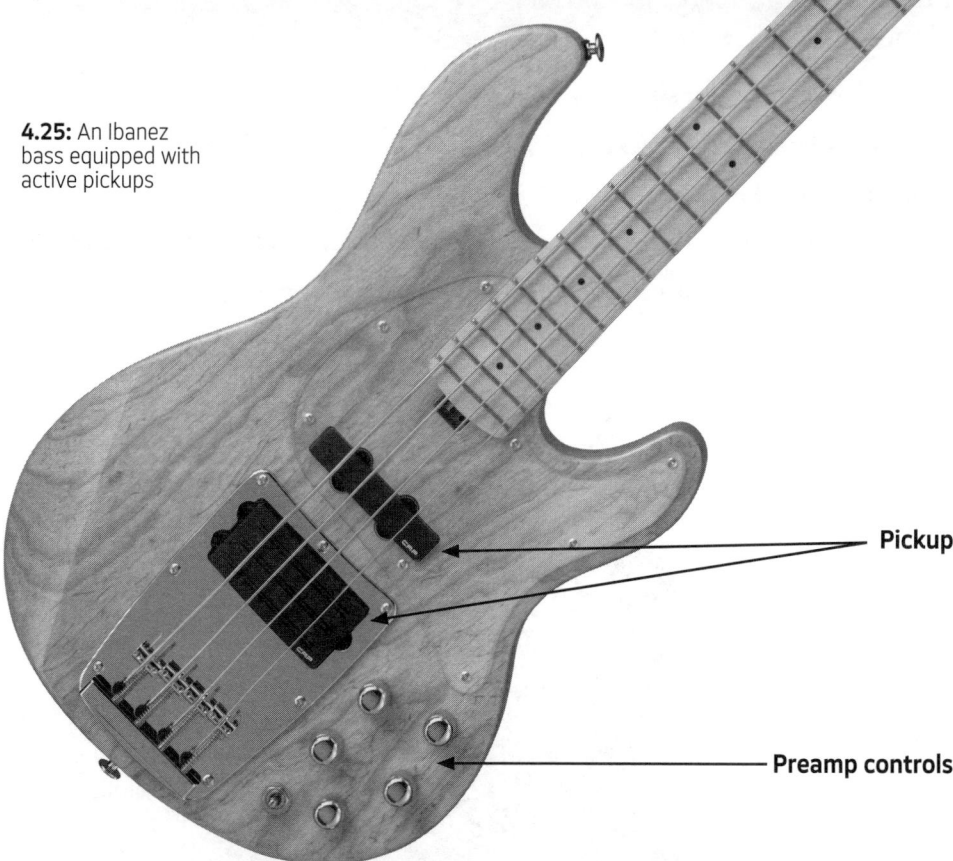

4.25: An Ibanez bass equipped with active pickups

Pickup

Preamp controls

and basses (Figure 4.25), and usually come in the same standard shapes used for magnetic single-coil and humbucking pickups.

CONTACT PICKUPS

Instead of focusing solely on the strings, contact pickups read the resonances and vibrations produced by the instrument itself. On stringed instruments, pickups are normally mounted in the bridge (under the saddle) or fixed to some part of the body (usually the top). Contact pickups can be added to an existing instrument. The term *acoustic-electric* refers to instruments that come with such a pickup already built in. The most popular contact pickups are the *piezoelectric* type used with acoustic guitars (Figure 4.26). Typically, the thin piezo element goes under the bridge saddle and feeds a preamp mounted on the body. The preamp—which requires a battery or phantom power to run—boosts the pickup's signal enough to feed an amp or mic preamp and usually offers volume and tone controls.

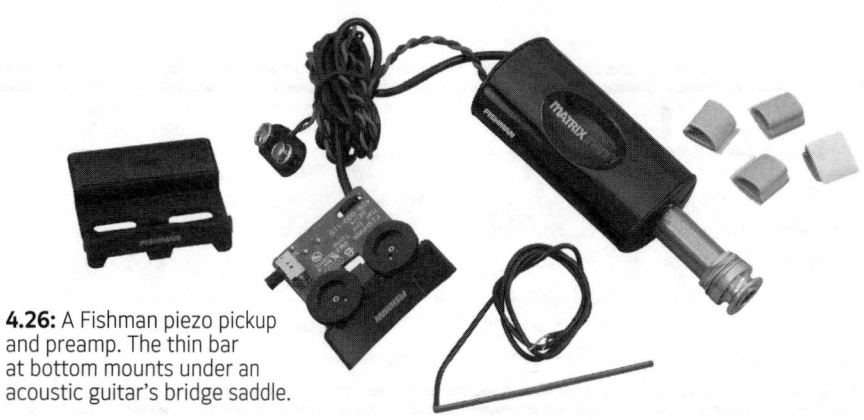

4.26: A Fishman piezo pickup and preamp. The thin bar at bottom mounts under an acoustic guitar's bridge saddle.

MULTI-CHANNEL PICKUPS

Multi-channel pickups can read the vibration of each string independently and generate a separate electrical signal for each one. Sometimes known as "hex pickups" because they can put out six separate channels from one guitar, multichannel pickups like Roland's GK-3A (Figure 4.27) allow guitarists to trigger sounds in synthesizers and other specialized gear, either directly or by converting the strings' vibrations into *MIDI messages* (see Appendix 2). With a hex pickup, each string can be processed to produce its own sound.

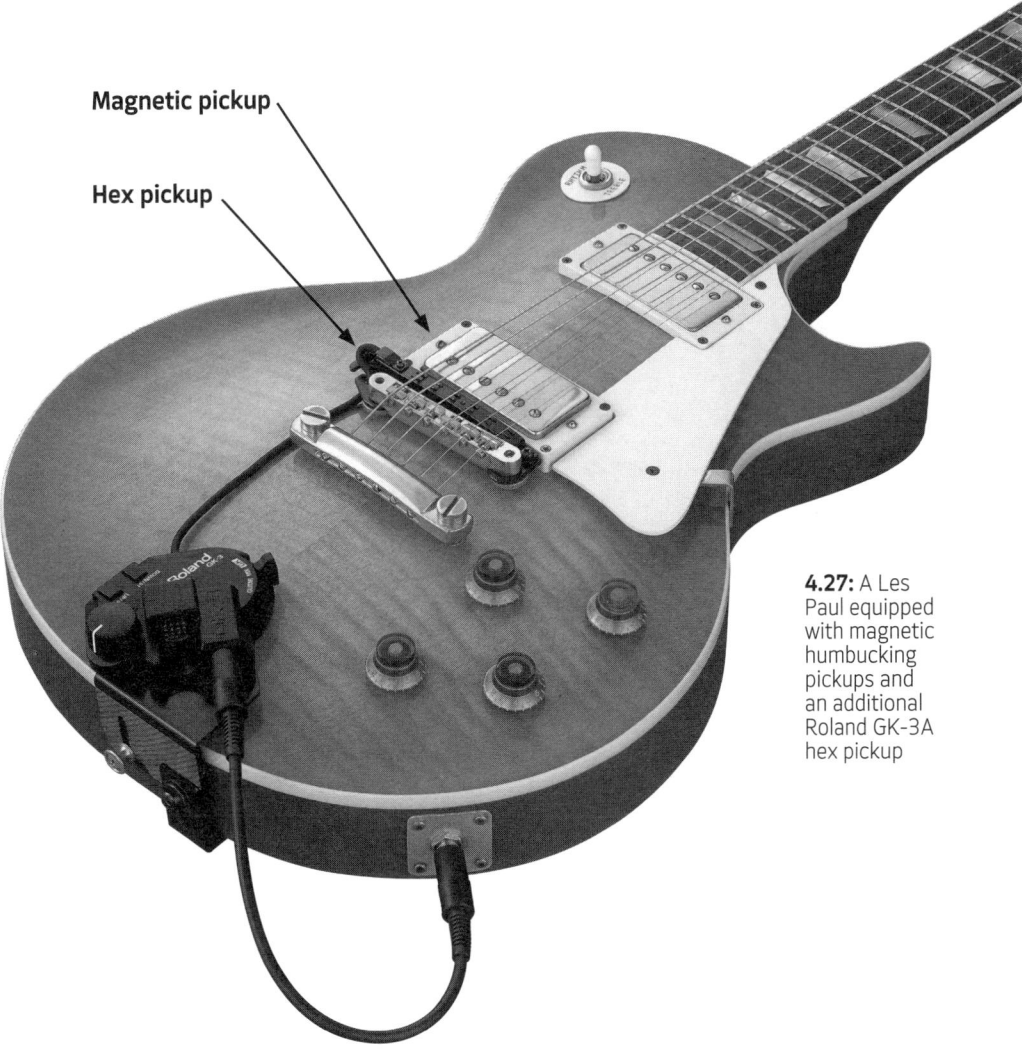

Magnetic pickup

Hex pickup

4.27: A Les Paul equipped with magnetic humbucking pickups and an additional Roland GK-3A hex pickup

Next Stages

With mics, pickups, amps, mixers, and speakers, you have everything you need to get sound from the stage to the audience. But you can improve and otherwise alter that sound with audio signal processors, which we'll explore in Chapter 5.

SIGNAL PROCESSORS AND EFFECTS

W hen you listen to a great-sounding band, what you're hearing is rarely just the result of some good-sounding instruments and clever microphone placement. Signal processors—also known as audio effects—play an important role as well.

In this chapter, we'll take a quick look at some of the most popular types of effects and explain their basic functions, but you should do more research to learn how these effects can be used on your instrument. First, let's start with the very basics.

What's a Signal Processor?

A signal processor is any device that can modify the sound of audio as it passes through your system. Technically speaking, mixers and amplifiers are signal processors. But in real-world use, the term usually refers to devices that are added to a system solely to modify or enhance the sound.

Many modern mixers—especially digital mixing boards—have signal processors built into them. These go beyond the equalizers that are typically found on mix channels to include things like compression, spatial effects, and more.

ADDING EFFECTS

Most mixers also have connections for external devices. On smaller mixers, these might be limited to one or two sends and returns, which are typically used for things like reverb and delay. On mid-sized and larger boards, there may also be channel inserts, allowing you to put signal processors directly into an individual channel. We discussed these different types of connections in Chapter 2, but you'll also see the same terms as we explore individual families of effects and their various applications.

Rack Effects

Usually, the types of signal processors that work best with a mixer are housed in rack units (they're often referred to as "rack effects," but the term says more about their size and shape than what they actually do). Figures 5.1-5.3 show a selection of rack processors.

Rack effects can use analog or digital signal processing. Most analog signal processors do one thing: for example, one will handle dynamics control, another equalization. Digital units often combine multiple effects, such as reverb, delay, modulation, dynamics, EQ, and more. These are called, appropriately enough, *multi-effects*.

5.1: Compressor/limiter.

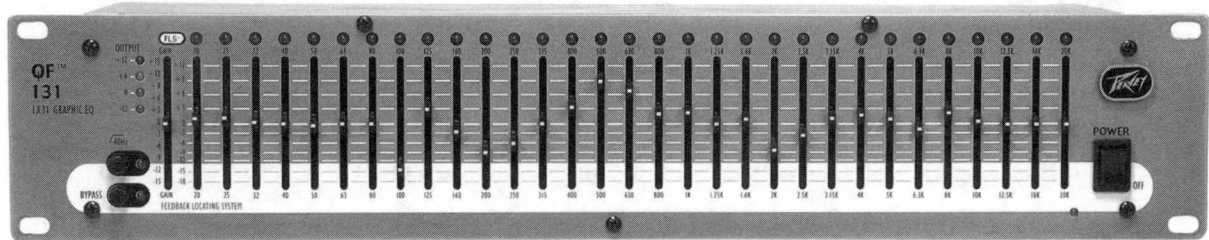

5.2: Graphic equalizer.

5.3: Multi-effect.

Pedals

Instrumentalists and individual vocalists can also use signal processors independent of the mixer. These come in two basic groups: *pedals* that sit on the floor, and rack effects like those we just mentioned.

Each category can include a huge array of options. Professionals often use a combination of pedals (also known as "stompboxes" because they look like boxes and people stomp on them to turn them on and off) and rack units. Some pedals—such as the wah-wah—are designed to allow the player to change the sound as he or she plays.

While most stompboxes handle one effect at a time, there are stompbox multi-effects. These are designed to offer the flexibility of a rack multi-effect, in a setup that's more compatible with typical stompbox uses.

As with a mixer, different effects might be placed early in the signal chain before an amplifier, or—if the amp is so equipped—in a connection called an *effects loop*. More on this later. Figures 5.4-5.7 show a variety of stompboxes.

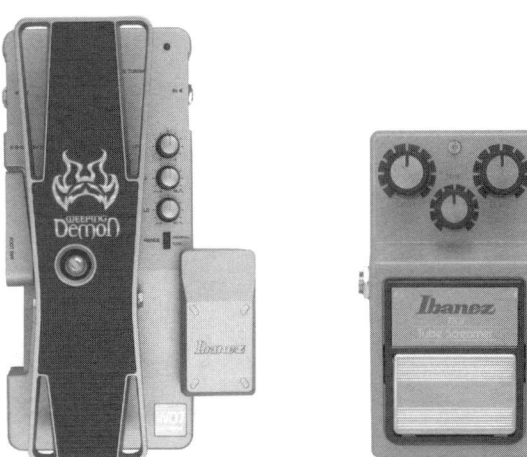

5.4: The motion of a wah-wah pedal (left) changes its sound.

5.5: The controls on this overdrive pedal (right) are set away from the on/off switch to prevent unwanted changes.

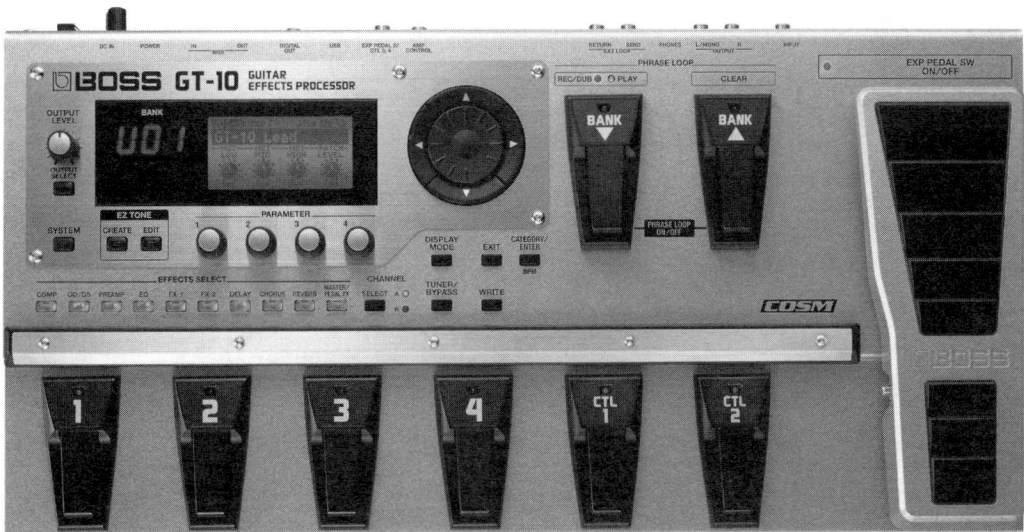

5.6: A stompbox multi-effect for guitarists.

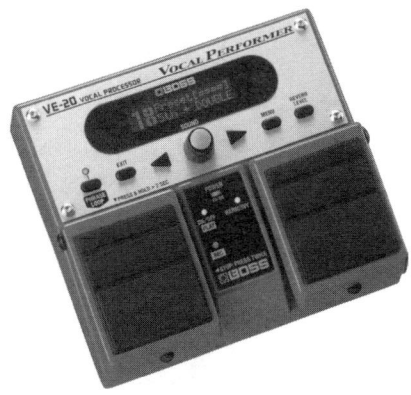

5.7: A stompbox-style processor for vocalists.

Built-In Effects and Computer Plug-Ins

As we mentioned at the beginning of this chapter, many digital mixers come with an assortment of built-in effects. So do amplifiers.

Vintage amps might have analog spring reverb and/or tremolo (sometimes mislabeled as "vibrato"). More modern amps can have digital reverb, delay, chorus, and more. Bass amps often have compressors built into their circuitry, and may include other effects as well.

Figures 5.8-5.10 show some examples of built-in effects.

5.8: A tube amp with reverb.

5.9: A bass preamp with a built-in compressor.

5.10: An amp with a digital effects unit built-in.

As a category, electronic keyboards may have the most impressive array of built-in effects. In fact, you'd be hard-pressed to find a multi-effect device that can outdo the internal effects found on a typical synthesizer or *workstation* keyboard like the one shown in Figure 5.11 (the kind of keyboard that includes piano, organ, strings, and other "acoustic" sounds).

5.11: Workstation keyboards typically have multi-effects processors built into their circuits.

Computers are playing more of a role onstage as controllers for things like electronic keyboards. They can even be used as mixers and guitar preamps! Audio inside a computer can also be modified with signal processors. These effects are called *plug-ins* and they come in a wide variety. Some try to recreate natural sounds like echo and ambience. Others, like the guitar processor shown in Figure 5.12, are designed to behave like pieces of audio hardware such as amps and stompboxes.

5.12: Peavey's ReValver software recreates amps, rack effects, and stompboxes.

Computers can also produce many of the same sounds you'll find on hardware synthesizers and workstations. But whether you're talking about hardware or software, stompboxes or racks, most signal processors fall into the basic categories listed below.

Types of Effects

It's easier to understand audio effects if you group them into basic categories. Although members of these different groups can often do similar things musicians and engineers usually separate them as follows: Gain boosters, dynamics processors, equalizers, modulation effects, pitch shifters, and ambient (or spatial) effects. Here's a quick rundown of what each group does and how it works.

GAIN BOOSTERS

Gain boosters do what their name implies: They increase the gain of an audio signal. They're most commonly used on electric guitar and bass, but can be used on anything—even vocasls or violins. Gain effects come in four main categories (Figure 5.13). These generally describe the sound of the effect, more than the way it works. Three of the four alter the sound by boosting gain of the incoming signal so that it causes overload, or *clipping* in the circuit inside the pedal. One knob determines how much clipping there'll be, another sets the overall output level. Some pedals also have EQ and other tone-shaping controls. Genrally, *overdrive* effects try to imitate the sound of a tube amplifier (some even use tube circuits). *Fuzz* produces a heavy, buzzing tone with lots of sustain (a great example is the sound on the Rolling Stones song "Satisfaction."). *Distortion* is somewhere in the middle, and the word is sometimes used as a general term for all three types. *Clean boost* adds level to the output to that the instrument pushes the amp's input harder, but the pedal itself doesn't produce any distortion.

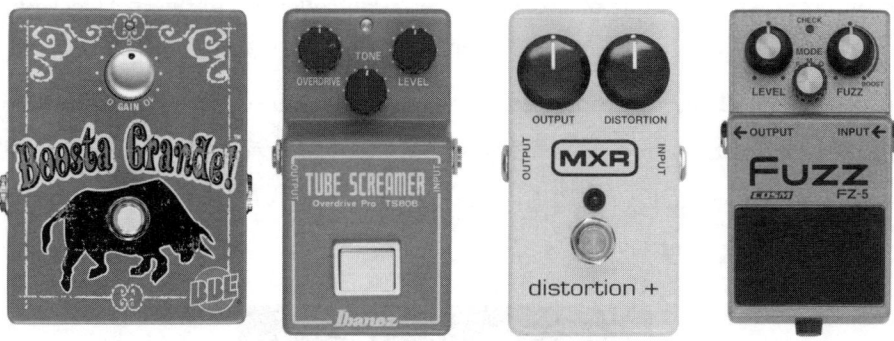

5.13: Four types of gain effects, in order of intesity from left: clean boost, overdrive, distortion, and fuzz.

DYNAMIC EFFECTS

Dynamics processors include compressors, limiters, gates, and expanders. They're usually used as *insert* effects to process one track at a time, or as *inline* effects in front of a guitar or bass amplifier. A dynamic effect's main job is to control the difference between quiet and loud signals—something known as *dynamic range.*

Compressors and Limiters

A compressor reduces dynamic range (the difference between loud and soft signals) by *attenuating* (making quieter) any signal that exceeds a certain level called the *threshold.*

The further the signal goes over the threshold, the more it's reduced. The amount of reduction is controlled by the *ratio.* A high ratio reduces the signal by a greater percentage. Some compressors have controls for attack and release, which determine how quickly the compressor acts when the signal goes over the threshold and how long the compression stays in effect when the signal drops below the threshold. Figure 5.14 shows the front panel of a compressor.

5.14: A compressor, with controls for input level, threshold, ratio, and output.

You might think that because they reduce signal levels, compressors are designed to make the overall sound quieter. But that's not really true. Compressors have an output level control that can actually boost the overall signal level. So in practice, they can actually *increase* the volume of quiet sounds. How?

Let me explain: A compressor only attenuates signal that's *above* the threshold. Everything below is left alone, and travels on its merry way towards the output. By reducing the loudest signals, you can turn up the overall level—the compressor will prevent the loudest signals from being *too* loud.

Compressors are used for vocals, drum mics, and other sources. Guitarists and bassists use compressors in front of their amps to help increase sustain. Compressors are also widely used for recording.

Limiting is a form of compression that stops signal from exceeding a specific level—no matter what! Limiters are typically used at the mixer's output to prevent signal from the mixer from overloading the power amps and damaging the speakers. Many compressors are called compressor/limiters because they include a separate limiting feature. Some power amps have built-in limiters.

Figure 5.15 shows the difference between compression and limiting. As you can see, a compressor merely reduces the signal level; a limiter completely stops it from exceeding the threshold.

5.15: Compression reduces signal above the threshold, but doesn't completely stop it the way limiting does.

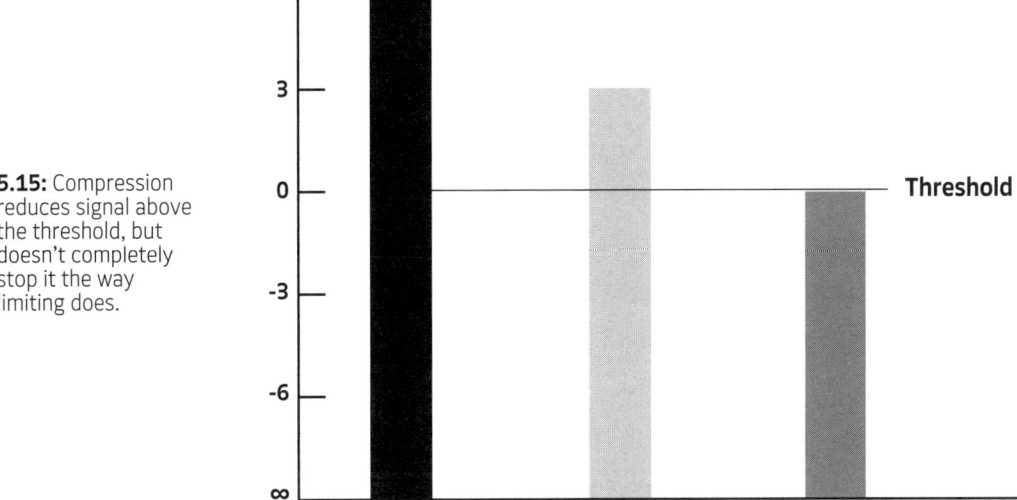

Onstage, compression can help control dynamics, but you should be careful not to use too high a ratio and too great a boost. You may actually increase noise—and cause feedback.

To prevent this, many compressors built for live sound have processors called *noise gates* built into them. We'll explain those next.

Gates and Expanders

While compressors and limiters reduce the level of signal *above* the threshold, a gate (or noise gate) mutes any signal that falls *below* the threshold. Gates can be used to eliminate unwanted noises that are low in signal level.

For example, by placing a gate on a vocal mic, you can automatically mute the mic when nobody's singing at it. This not only prevents bleed and feedback but can help reduce rumble coming from the mic stand and other unwanted noises.

Gates are also used to prevent hiss and other noises introduced by electronics. For example, a guitarist might use a gate so that there's no hum from his pickups when he's not playing. Figure 5.16 shows how a gate works.

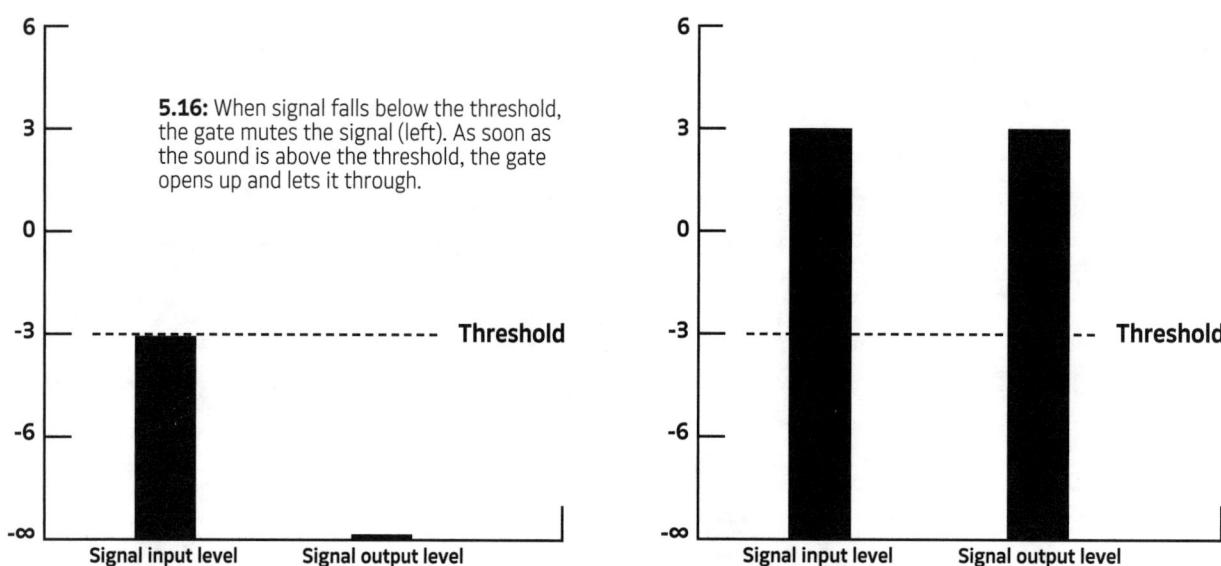

5.16: When signal falls below the threshold, the gate mutes the signal (left). As soon as the sound is above the threshold, the gate opens up and lets it through.

Although it can reduce noise, a gate isn't really a *noise reduction* device in the strictest sense. That's because when the gate opens, every part of the sound comes through. For example, the guitar pickup hum we mentioned earlier *will* be part of the signal when the guitarist is playing. But because he's also going to be producing sound, the pickup hum won't be noticeable. It's only when the guitar is quiet that you'd hear it. The gate is perfect for preventing that kind of problem.

Setting the gate threshold requires care. If it's too high, it might not pick up a singer's quiet passages; if it's too low, it'll pick up too much unwanted noise.

Gating is an extreme form of *expansion*—which increases the dynamic range by reducing the level of any signal that falls below the threshold, while leaving signal above the threshold intact. You can think of a gate as the reverse of a limiter, and an expander as the alter ego of a compressor.

EQUALIZATION AND FILTER EFFECTS

Equalization is a fancy way of saying "tone control." You'll find many types of equalization, or *EQ*, in live sound. We already addressed this in Chapter 2. *Filtering* is another way of saying EQ.

The most common types are:

Graphic EQ

On a graphic equalizer, a slider controls each area of the frequency spectrum. With a standard analog graphic EQ, the frequencies are fixed, but on digital EQs, you can often set the bands to any value that fits the current application. A rackmounted graphic EQ is pictured a few pages back, in Figure 5.2. Figure 5.17 shows a similar effect in stompbox form.

5.17: A stompbox graphic EQ

Shelving, Low-Pass, High-Pass, and Band-Pass Filters

Shelving EQ affects signal above or below a specific frequency, or *hinge point*. Since these controls are commonly found on mixer channels, we already explored them in Chapter 2.

A *low-pass* (or *hi-cut) filter* attenuates frequencies *above* the specified hinge point. Lower frequencies pass through—hence the name.

A *high-pass* (or *lo-cut) filter* attenuates frequencies *below* the hinge point. High-pass filters—sometimes seen in the form of "low cut" switches—are especially useful onstage because they can reduce rumble coming into mics.

They can also prevent *resonant feedback*—a condition that occurs when a certain frequency causes the body of an instrument to vibrate uncontrollably. This is pretty common with acoustic guitars. Notch filters (see below) can also help tame resonant feedback.

Believe it or not, some mics have high-pass filters built into them—little switches on the mic's body that can reduce stage rumble.

A *band-pass filter* processes frequencies above and below the specified frequency range.

Parametric EQ

Each band of a parametric EQ (Figure 5.18) gives you control over three parameters: center frequency, bandwidth, and boost/cut level. Bandwidth or **"Q"** determines how frequencies near the center are affected.

5.18: Parametric equalizer.

Wider bandwidth affects a larger frequency range. A narrow bandwidth focuses more closely on the center frequency. A very narrow-bandwidth EQ is sometimes called a *notch filter.*

Notch Filters and Feedback Eliminators

While most EQs are designed to enhance a sound in a noticeable way, notch filters have a stealthier job. By focusing on a very narrow part of the sound spectrum, they can prevent problems without changing the sound so much that the listener notices.

That's one reason why notch filters are often included on acoustic guitar preamps (Figure 5.19). When the guitar starts to howl, sliding the notch filter can help tame the offending frequency without changing the guitar's overall tone very much.

5.19: The notch filter on an acoustic guitar preamp.

A set of notch filters can be especially effective at reducing feedback on an overall mix, using a technique called "ringing out the room." Specific frequencies causing feedback are identified using a tone generator and notched out of the signal. Automatic feedback eliminators (Figure 5.20) can monitor feedback and set the appropriate notch filters. These are designed to go between the mixer and power amp as a last stage of processing.

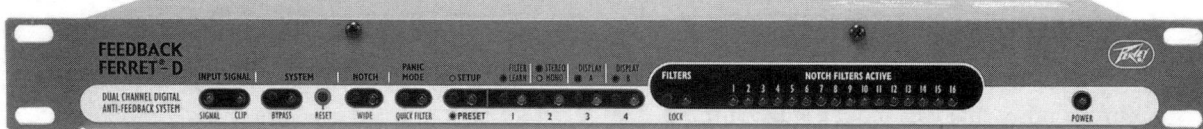

5.20: A feedback eliminator

MODULATION EFFECTS

Modulation effects include chorus, flanger, phaser, tremolo, and vibrato. Modulation effects are used for sweetening—adding a little flavor to an otherwise static sound.

The word *modulate* means to change. You may have encountered the term in a musical context when a song changes key, for example. In audio, modulation usually refers to signal processors that change the sound in a repeating cycle that takes a period of time to run its course.

The easiest way to understand this is by looking at a tremolo effect (Figure 5.21). Tremolo is a steady change in volume, where the signal goes from a starting level to a lower volume and back up again in rapid fashion. The amount of time it takes for the volume to reach its low point and return to the starting point is one cycle.

5.21: Tremolo in action

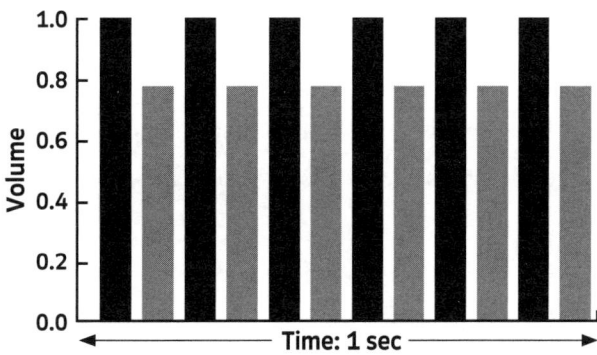

The number of times the cycle runs its course in one second is known as the *frequency* (also called the speed or rate) of the effect. The amount of change that occurs during each cycle is known as the *depth*, or intensity, of the effect. Figure 5.22 shows how differences in speed and depth affect a tremolo sound.

5.22: Changes in the speed and depth of a tremolo effect can make it more pronounced or more subtle.

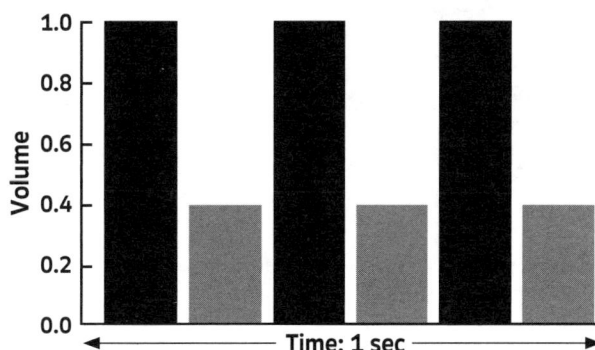

Here's a brief rundown of the most popular modulation effects and their applications.

Tremolo and Vibrato

Tremolo and vibrato are the two most basic modulation effects. Tremolo affects volume; vibrato affects pitch. Tremolo effects are found on vintage guitar amps, in stompboxes, and in multi-effects. Fender amp controls that say "vibrato" are actually producing a tremolo effect. (Ironically, the vibrato bridge on Fender's Stratocaster is called a tremolo, a misnomer that has stuck to this day!)

Vibrato effects can be found in stompboxes and rack effects, though some amps do have actual—that is pitch-modulating—vibrato built into them too.

Chorus

Chorus effects are designed to make a single instrument sound like it's part of an ensemble. Well, at least that was the idea when they were invented. In reality, they add a shimmering texture that can make an instrument sound lush, or a tremble that can make it stand out in other ways.

Chorus generally works by modulating pitch very slightly and/or delaying the signal by a small amount, then sweeping from a neutral pitch to the changed pitch and back.

Chorus is typically used on electric and acoustic guitar, keyboards (especially electric piano), and occasionally on bass and vocals. Figure 5.23 shows some typical chorus settings. You'll find chorus effects in rack processors, some mixers, and in a dizzying array of stompboxes.

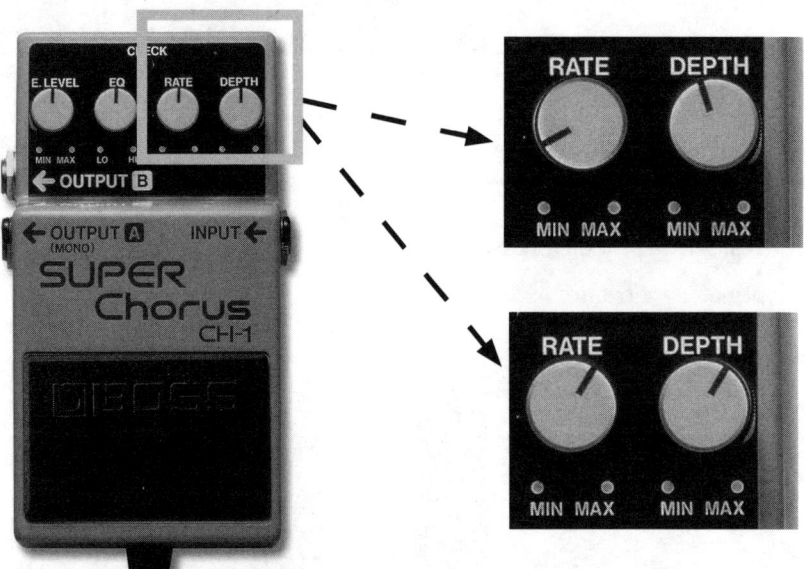

5.23: Chorus settings: Slow rate/low depth adds some subtle color to rhythm guitar (top). Faster rates and higher depth creates a vibrato-like effect (bottom).

Effects like the Uni-Vibe (Figure 5.24) are also within the chorus family. The Uni-Vibe and similar devices use a pedal to control the speed of the modulation; this actually simulates the action of a rotating loudspeaker, like those found in the famous Leslie cabinet shown in Figure 5.25. The Leslie used moving speakers mounted within its cabinet and was an important part of the Hammond organ sound. Because the cabinets are so massive and heavy, Leslie effects are usually recreated electronically today.

5.24: Uni-Vibe.

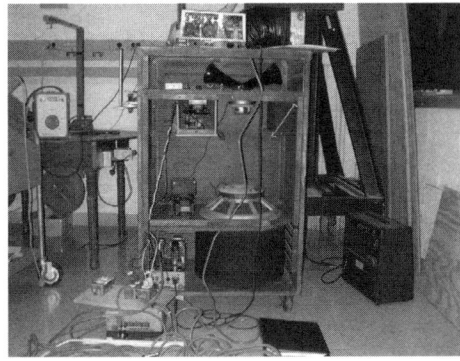

5.25: Leslie cabinet.

Flanger

Chorus and flanging are so closely related that some stompboxes actually do both. Like chorus, flangers create a sweeping, shimmering sound. But a flanger (Figure 5.26) uses delay and filtering to create its effect. Most flangers also have a *regeneration* control, which feeds a little of the effected sound back into the mix to further intensify the effect. This produces some pretty exotic sounds. One of the most famous is the "jet engine."

5.26: Flanger.

Flangers are most commonly used on electric guitars and keyboards—they're a little too metallic-sounding for typical acoustic guitar duties. A flanger set to a big sweep with low regeneration can be especially cool with distortion. You have to experiment with the controls a little bit, but it's worth it.

Phaser

Phasers sound similar to chorus effects, but they work by changing the frequency balance in a way that creates what's known as a *comb-filtering* effect. (If you track on a chart the frequency response of a phased audio signal, it looks like the teeth of a comb.) Phasers are used in almost the same settings as chorus, though they tend to have a thicker sound.

PITCH SHIFTERS

Although they're sometimes grouped with modulation effects, we're going to look at pitch effects—also known as *harmonizers*—as their own category. They come in several varieties, but they all work by changing the pitch of an incoming signal.

Basic Pitch Shifters and Octave Dividers

These devices raise or lower the pitch of a signal by a fixed amount. On a pitch shifter, this can be any interval—for example, two semitones higher or five semitones lower. A pitch shifter can also change a signal by a small percentage or *cent* of a semitone. (A cent is 1/100 of a semitone—so an octave is 1,200 cents!) This creates an interesting chorus-like effect.

Some pitch shifters are designed to produce a note an octave higher or an octave lower than the source pitch. These are known as *octave dividers*. You'll find them both as stand-alone devices or in combination with distortion, as in the famous octave-plus-fuzz effects used by Jimi Hendrix. Figure 5.27 shows an octave fuzz.

5.27: Octave fuzz combines distortion with pitch shifting

Octave effects are popular on electric guitar and can also be used by bassists (or in the bass or kick-drum channel of a mixer) to add low end to the sound.

Intelligent Pitch Shifters

If you mix a pitch-shifted signal with the original sound, you get a harmony, but that harmony won't always play the correct note—at least not if you plan to play in key. Check out Figure 5.28: it shows what happens when you set a pitch shifter to four semitones (a major third) and play a C major scale. The first note sounds great, but the second, third, sixth, and seventh are all *outside* the key.

5.28: The harmony produced by a fixed pitch shifter adds notes that are not in the key of C (those with sharp signs).

To play the harmony and still be in the correct key, you'd need to have something like Figure 5.29; you'll notice that in some places, the harmony note is four semitones higher than the played note, while in others it's only three semitones higher. An intelligent pitch shifter can make such an adjustment; you tell it what key and interval you want, and it produces the harmony.

5.29: An intelligent pitch shifter produces a harmony that stays in the key of C.

Whammy Pedals

Like chorus pedals and filters, pitch effects can also be controlled in real time. Perhaps the most common example is the Whammy pedal (Figure 5.30), which lets a player change pitch with a foot pedal. Whammy pedals are most popular among rock guitarists. Keyboard players can also use Whammy pedals, but they probably don't need to—the pitch wheel (or joystick) on their instrument can accomplish the same thing.

5.30: Whammy Pedal

Pitch Correction

One of the hottest effects in the recording studio, pitch correction is also used in live situations to help singers avoid those embarrassing off notes. Additionally, it's used as a special effect to produce the "T-Pain" stairstep robot sound. You'll find it built into rack processors and floor effects designed for vocalists (Figure 5.31), as well as in software plug-ins.

5.31: A vocalist's stompbox

Pitch correction works by sensing the incoming pitch and automatically adjusting it to meet an "in tune" note. It can work chromatically, or be focused on a specific key. Personally, I don't think it's a great idea to use pitch correction when you're performing, especially if you're just learning your craft. Part of the process is learning to sing on pitch consistently.

However, if you do decide to use pitch correction, it's essential that the corrected signal does *not* go to the singer's monitors. If it does, he or she will keep adjusting to the corrected pitch and actually sing even more out of tune!

DELAY AND ECHO

Delay is among the most popular effects on every instrument—it can even be used on percussion. In live sound, it has both creative and problem-solving applications. We'll spend more time on the creative side.

You can use delay to create doubling effects (a short delay can make it sound like more than one instrument is playing), slapback sounds reminiscent of early rock & roll, and long, repeating echoes. Combined with modulation effects, delay can be used to paint lush audio landscapes.

Parameters include *delay time*, which controls the amount of time between the initial signal and the repeats; *feedback*, which controls the number of repeats; and *mix* (see the sidebar on the opposite page). Delay time can be set to a fixed number of milliseconds (1000ms = 1 second) or can be set to a musical subdivision, such as a quarter note, and synced to a tempo. In the latter case, you can use your foot to "tap in" the tempo by pressing the pedal in the appropriate rhythm. This feature is known somewhat cryptically as *tap tempo*. Figure 5.32 shows a delay pedal with a control for tap tempo.

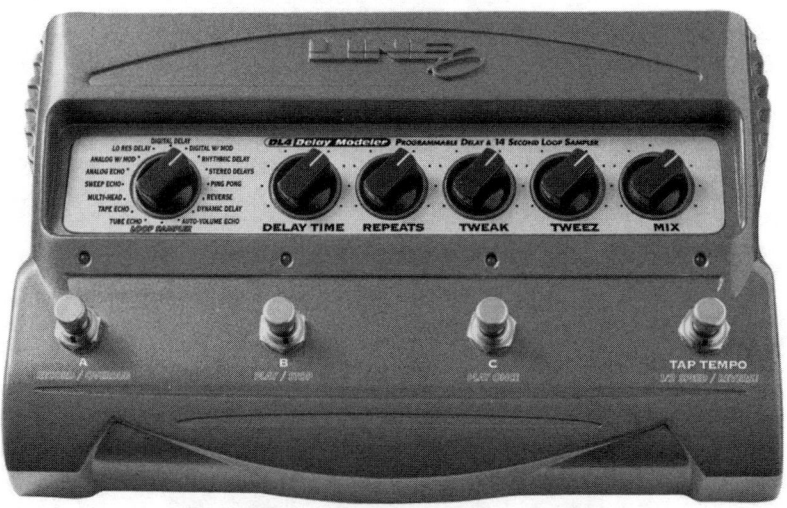

5:32: To use a delay pedal with tap tempo control, press the tap button once to define the beginning of the beat. Press it a second time to define the end. If, for example, the two taps are 1 second apart, an eighth note-delay would repeat every half-second (500 milliseconds).

Mixing on the Box

Some modulation and most delay and reverb effects have a parameter we haven't discussed before: Mix. Why?

Well, remember when we talked about sends and returns in Chapter 2? A mixer's send/return system lets you blend in just the amount of effect you want. This way, you can set all effects devices connected to the return to 100% effect and use the mixer's knobs to control the balance.

But what if you want to use an effect in line? This might happen if you plugged the effect into a mixer channel's insert. Stompbox effects are also typically arranged in line—there are examples of send-and-return setups for pedals, but they're the exception and not the rule.

Effects like compressors, gates, EQ/filters, and distortion boxes are designed to be mixed at 100% all of the time, so they never have mix controls (they do have "level" and "gain" controls, but these govern the overall effect, not the way the processed sound blends with the unprocessed, or "dry," signal).

But modulation, delay, and reverb effects often sound better when only a small amount of the effect is blended with the original signal. If you were to run a delay pedal at 100% mix on, say, a guitar, all you'd hear would be echoes—you'd never hear the original note. The mix knob allows you to set this parameter to a better balance (usually 50% or lower).

Delays come in multi-effects units, may be built into some mixers and amps, and can be found in a variety of stompbox configurations, including those that can switch between several delay modes. Delays can use analog or digital circuitry. Many digital devices are capable of producing several different varieties of delay and echo effects:

Mono delay sends the original and delayed signals through the same output.

Stereo delay lets you set separate delay times for each stereo channel.

Ping-pong delay bounces the delayed signal between left and right channels.

Panning delay gradually pans the delay across the stereo spectrum.

Tape delay and *modulation delay* process the delayed signal to simulate the sound of a tape-based delays and other modified delay sounds.

REVERB

Reverberation—or *reverb* for short—is probably the most widely used effect of all. It creates a dimensional space around the input signal. Everybody seems to like the way they sound better when their voice or instrument has some reverb on it.

In the studio, reverb is almost a given. But on stage, it can be more problematic. First, the room you're playing in may be very dry-sounding, or "dead." Or it may be very echoey, or "live." In the latter case, the room is already creating natural reverb on the sounds you're making. Adding more can cause things to sound distant, hollow, or otherwise bad.

But reverb can also be part of a player's sound. Guitarists who use reverb-equipped tube amps often say that their overall tone changes when the reverb is turned off because it's actually part of the amp's main signal path. My recommendation is to use less reverb on stage than you would in the studio and see if you can do without it.

Reverb Types

Reverb units used in live performance come in two forms. *Spring reverbs* are physical devices that use springs in a small metal tank to create ambience. These are found built-into guitar amplifiers and are also available as standalone units.

Much more common are *digital reverbs* (Figure 5.33), which use software *algorithms* to imitate the ambience produced by physical spaces (like concert halls, rooms, and chambers). Digital reverbs may also imitate the spring reverbs mentioned above, as well as a type of artificial reverb used in recording studios called *plate* reverb.

5.33: Compact reverb pedal

Depending on the model and the type of reverb effect you're trying to create, you'll find pretty wide variation in the kinds of controls at your disposal. However, most digital reverbs offer at least some of the parameters below:

Pre-delay controls the amount of time it takes before the reverberant signal is heard.

Reverb time determines the length of the reverb's decay.

Early reflections set the delay time and loudness of the first echoes in a reverberant sound.

Gate applies a noise gate to the reverb's decay and can be used to make the reverberations cut off abruptly.

EQ or frequency adjust determines how the reverb will respond to specific frequencies.

Mix determines the balance between the reverberations and the original, or dry, signal.

Delay to the Rescue!

As I mentioned earlier, delay is also used to solve sound reinforcement problems. It's unlikely that you'll be dealing with these situations on your own, but you should know that when speakers are very far apart and you're standing much closer to one of them, you can hear a lag between the two speakers—that's the time it takes for sound to travel from the distant speaker to your ear. This can change the phase of the audio as it reaches the listener. To help solve this, engineers might use a small amount of delay to "time align" the speakers. You'll see this feature built into some digital mixers, and now you know why!

Digital reverbs may be built into mixers and instrument amps, but you can also add them to your system rack effects (reverb is a big feature on most multi-effect devices), and as stompboxes.

Reverbs are typically used as send/return effects, with just a small amount of the reverb mixed in to create a sense of space around the dry signal. Too much reverb can drown a sound (maybe that's why they call it wet!). Therefore, if you plan to use a reverb effect in line with your main signal, you should probably set the reverb's mix control pretty low. It should usually go last in the signal chain.

The Order of Effects

The order in which you connect audio effects can make a big difference to their sound. Effects like gain boost, compression, and EQ (including wah pedals) usually go first in the signal chain. Depending on the instrument and the sound you're looking for, you might swap them around. If EQ feeds a gain effect, for example, it can boost certain frequencies going into the gain unit and influence the character of the distortion. Placing it after the distortion will also work, but may sound quite different. But you can decide which you like better.

Next come the pitch and modulation effects like chorus, delay, phaser, etc., followed by echo/delay, and finally reverb. But there's no rule that says you have use the same order. Try running a delay into a distortion box and see what you think. You might like it! If your amplifier has an effects loop, you might want to use gain, compression, and EQ between the instrument and the amp, and run modulation, delay, and reverb through the loop.

Effects can go in series—so that the output of each one feeds the next—or they can run in parallel, where two or more effects get signal from the same source, but feed the outputs independently. A mixer's insert is an example of a series effect; the send/return system is an example of a parallel effect.

Next Stages

Now that you have a basic understanding of signal processors and effects to go with your knowledge of mixers, amps, microphones, and speakers, it's time to start cabling things together. That's what's coming in Chapter 6.

Chapter 6

MAKING CONNECTIONS

So far, we've talked about all the major pieces of a sound reinforcement system: mixers, amps, speakers, microphones, and effects. That's a lot of stuff! But it doesn't matter how cool your instruments are, how awesome your amps and effects sound, or how great your mics are if the cables that connect the components don't work. How do you connect them all together?

A complete sound reinforcement system can include lots of different wires and connectors. There will be connectors for audio and for electrical power. There may also be connectors for computers and MIDI instruments, such as keyboards and electronic drums. Some of the audio may even be traveling to and from the sound reinforcement system via wireless transmitters.

We'll talk about how and where to use these various connectors in much more detail in the second half of this book. For now, let's just learn to identify various cables and connectors and concentrate on their basic uses.

Analog Audio Cables

Audio cables come in a few common categories: microphone cables, instrument cables, patch cables, data cables, digital audio cables, and speaker cables. As we discussed in Chapter 1, analog audio cables can be balanced or unbalanced. Digital audio cables also come in several varieties.

Before going into detail about different types of cables, let's take a second to look at the parts of a cable. That's right, cables have parts! They're not just wires.

The outside of the cable—the rubbery part you handle during day-to-day use—is called the *case*. Inside the case, there may be a layer of metal called the *shield*. Inside the shield are the *leads*. You'll usually find either two or three of these leads. This is important because the way the leads are wired to various connectors can have a big impact on the way the cable works and sounds.

Cables come in a variety of lengths and thicknesses. The most common lengths for microphones and instruments range between 15 and 20 feet. Longer cables are available, but are not recommended for instruments because the increase in length can affect the sound by reducing high frequencies. Short cables (ranging in size from about 6" to 24") are usually referred to as patch cords, or patch cables. The thickness, or *gauge*, of a cable can also affect its performance.

The following illustrations show a number of different cable types and their uses. Depending on the connectors used, these cables can be used in a number of different ways.

UNBALANCED CABLES

Unbalanced cables are used to plug instruments into amplifiers and connect effects on a pedalboard. Sometimes they're also used for microphones, or to connect mixers to amplifiers. They have two leads inside their cases. One, called the *hot* wire, is for the instrument's signal; the other wire, called the *ground*, limits the buildup of static electricity in a circuit. (Wires that aren't grounded can cause many problems, from noise to voltage surges.) Unbalanced cables are often connected to 1/4" mono plugs and jacks, as well as RCA connectors.

Figure 6.1 shows the wires inside an unbalanced audio cable.

6.1: Unbalanced cable showing the hot and ground wires — **Hot** / **Ground**

Unbalanced cables can also be used to connect power amps to speakers. Speaker cables are a little different from instrument cables; they often use heavier wires and don't need to be shielded. Instead of hot and ground, the strands of the cable are known as positive (+) and negative (-). Figure 6.2 shows a speaker cable.

6.2: Speaker cable — **+** / **−**

BALANCED CABLES

Balanced cables have three internal wires, as shown in Figure 6.3. The extra wire, a *cold* (or *drain*) wire, operates at the reverse polarity of the hot wire. This "balances" the signal, which helps reject noise and improve sound over long cable runs. Both hot and cold wires carry audio.

6.3: Mono balanced audio cable

All three wires inside a balanced cable (hot, cold, and ground) are connected to the jack or plug being used to connect the cable; it's very important that the connections on both sides of the cable match up. Balanced cables are typically used for microphones, but they can also connect two pieces of audio equipment—for example, a mixer to a power amplifier or set of powered monitors.

Unbalanced and balanced cables can be employed in many different ways, depending on the connectors on either end. Let's take a look at common connectors before moving on to various ways of using cables.

Basic Audio Connectors

Audio connectors come in several different families, and each family has two members: the male or *plug* end, and the female or *jack* end. Figure 6.4 shows an illustration of common connectors used for analog audio.

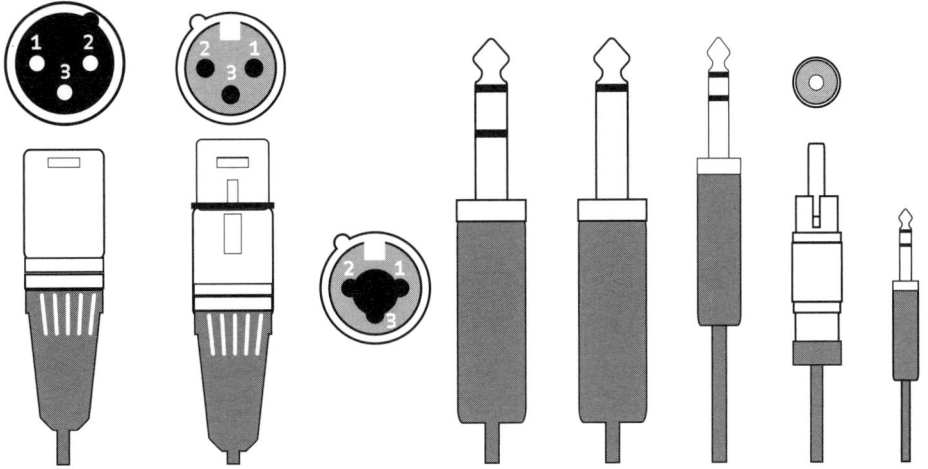

6.4: (From left) XLR male, XLR female, XLR/1/4" combo jack, 1/4" TRS, 1/4" TS, Bantam, RCA, 1/8" mini

XLR CONNECTORS

XLR connectors are the most common connectors used for microphones with balanced cables. They're also used to connect pieces of gear, such as mixers and power amplifiers. Professionals love XLRs—and for good reason: XLRs are sturdy, they lock into place (so they don't come loose by accident on stage), and they have two separate connectors—female and male—so you always know which end goes where. An XLR connector has three pins (Figure 6.5); usually, the hot wire runs to pin 2, the cold wire goes to pin 3, and the ground goes to pin 1. Both the male and female sides of the cable need to be wired the same way. If the hot and cold wires are reversed, the signal will be out of phase.

One thing that's cool about an XLR cable is that it's just like the jacks you plug it into: It has male and female ends, so you always know which end is which.

6.5: Typical XLR wiring

XLRs and Microphones

With an XLR, you must plug the mic into the female end of the cable. You then plug the male end into the mixer. The XLR's snapping feature helps hold the cable in place. Figure 6.6 shows an XLR connector snapped into a mic. As you can see, the cable can support the mic's weight. I don't recommend you carry your mics around by the cables like this, but it shows how well locking XLRs work!

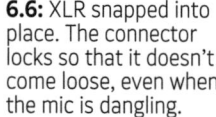

6.6: XLR snapped into place. The connector locks so that it doesn't come loose, even when the mic is dangling.

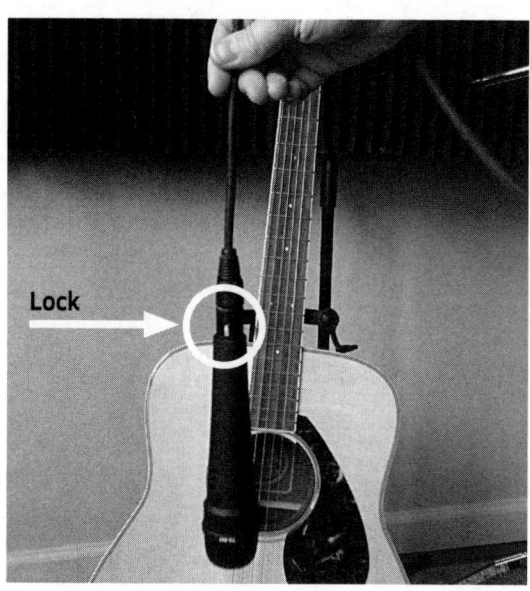

XLRs and Other Devices

XLR connectors are also used to make balanced *line-level* connections between audio devices. The most common example in sound reinforcement is between the mixer and the power amplifier (or powered monitors).

XLRs can also be used to run digital signals in the AES/EBU format.

QUARTER-INCH CONNECTORS

Quarter-inch connectors are the most common connector used to connect electronic instruments to amps and effects. They come in two types:

1. **Mono**, also known as tip/sleeve (TS) connectors because they have two connections: the tip (for the hot wire) and the sleeve (for ground)

2. **Stereo**, also known as tip/ring/sleeve (TRS) because they have three connections: the tip (hot), right (cold), and the sleeve (ground)

Figure 6.7 shows how the two 1/4" connectors compare. Whether they're mono or stereo, 1/4" cables typically have male plugs on both ends; the female jacks for both mono and stereo look identical from the outside, so it's important to check your gear to know what kind of connection is present.

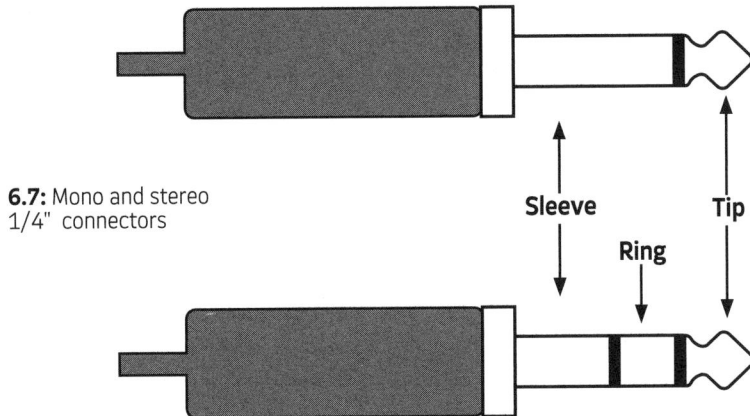

6.7: Mono and stereo 1/4" connectors

Sleeve   Ring   Tip

You'll commonly find 1/4" plugs in both straight and 90-degree form. The 90-degree connectors are especially handy when space is at a premium, such as on an effects pedalboard. Both patch cords and cordless connectors called "couplers" use 90-degree plugs.

Stereo TRS 1/4" connectors are used to connect balanced line-level devices, and, in some cases, route a stereo signal to a stereo input. They're also used to connect stereo headphones to the headphones outputs found on mixers, preamps, electronic keyboards, and so on. You can use a 1/4" TRS connector on one side of a cable and an XLR on the other. Typically, the tip would go to pin 2 (hot), the ring to pin 3 (cold), and the sleeve to pin 1 (ground).

Cable Configurations

So far, we've looked at each cable carrying either a balanced or unbalanced signal in one direction—for example, from a mic to a mixer channel. But by using different connectors and combining cables in different ways, you can route signals in other ways. Here are some of the more common examples.

BALANCED CABLE'S UNBALANCED USES

Becuase it has three leads, a balanced cable can also be used to run two separate *unbalanced* signals. For instance, it can run a set of stereo headphones, split a stereo signal into two mono signals, run the left and right sides of a stereo output, and route effects to and from a mixer or amplifier.

Stereo to Stereo

When running an unbalanced cable in stereo, it's most common to use a 1/4" TRS connector. The tip is the left side of the stereo signal; the ring is the right; the sleeve is the common ground (Figure 6.8).

6.8: A stereo cable with a TRS connector handling left and right signals

When two male connectors are used, the cable can plug a stereo output into a stereo input. When a female TRS jack is used on one end, the cable can be used as an extension cable for headphones.

Split Left and Right

Another commonly seen configuration is one unbalanced cable with separate connectors for the left and right sides. These can be created by bundling together two unbalanced cables, or by sharing the ground in a balanced cable.

Figure 6.9 shows an example with two sets of 1/4" plugs on each end. Similar configurations with RCA plugs, or a mix of 1/4" and XLR plugs, are also common.

6.9: A split stereo cable

These cables can be used two ways:

1. To route a stereo output into two mono inputs (Figure 6.10)
2. To route a mono effects send on one side and an effects return on the other (Figure 6.11)

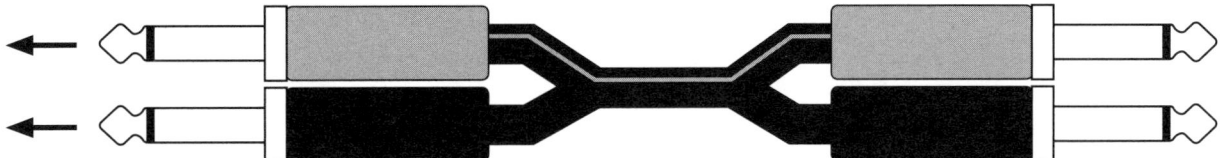

6.10: Using each side of a stereo cable to feed a separate mono input

6.11: Using one side of the stereo cable as an effects send and the other as an effects return

Stereo to Dual Mono

You'll also find stereo cables wired with a single stereo connector (such as a 1/4" TRS plug) on one side and two mono connectors on the other. This can be used two ways:

1. To split a stereo signal into two mono channels (Figure 6.12)
2. To run an effects send/return using a single jack on a mixer or amp (Figure 6.13). This is the common routing scheme for channel inserts, which we discussed in Chapter 2.

6.12: Using each side of a stereo cable to feed a separate mono input

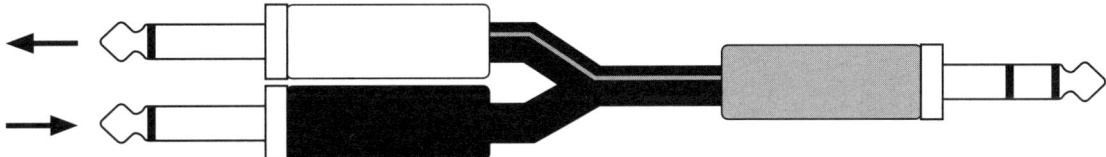

6.13: Using a single jack as an effects send and return

RCA, MINI, AND BANTAM CONNECTORS

While XLR and 1/4" connectors will probably be found on the majority of your gear, you'll also encounter a number of other audio connectors.

RCA Connectors

RCA connectors, also known as *phono* connectors, are commonly used to connect devices like tape and CD players to your audio system. Many mixers will include one stereo set of RCA connectors. You may also find RCA connectors on a portable audio interface or on a digital keyboard that's primarily designed for home use.

RCA connectors are also used to send digital signals in the S/PDIF format.

Mini Jacks

Stereo mini connectors—also known as 1/8" or 3.5mm plugs and jacks—are used for the headphones outputs of devices such as iPods and other portable music players. When used in stereo, these connectors use the same kind of TRS scheme used by balanced cables: Tip is one side of the stereo signal; ring is the other; sleeve is shared ground.

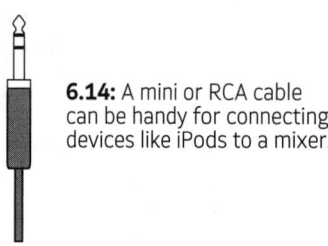

6.14: A mini or RCA cable can be handy for connecting devices like iPods to a mixer.

Few mixers have inputs of this type, so you may need to get a cable with a stereo male mini jack on one end and a pair of mono 1/4" or RCA jacks on the other, as shown in Figure 6.14.

TT or Bantam Connectors

These look like mini 1/4" TRS cables, and follow the same wiring scheme. They're usually used for patching gear together in professional patch bays (see the section on patch bays later in this chapter), but aren't that common onstage unless you're talking about a fully professional setup.

SNAKES AND MULTI-PAIR CABLES

To save clutter, groups of mono or stereo cables may be bundled together in one case. When they're all set up with connectors, these bundles are usually referred to as *snakes*. Each cable inside the snake is called a *channel*, and you'll typically see snakes with between eight and 24 connectors, as shown in Figure 6.15.

6.15: A 24-channel analog audio snake

Digital Audio, MIDI, and Data Cables

Digital cables that are designed to carry a stereo signal look a lot like analog audio cables, and you can use the same kinds of mono and stereo wires to connect some digital formats. However, cables that are designed specifically for digital audio are built to produce a cleaner signal, which can prevent some data errors.

MULTICHANNEL DIGITAL CABLES

There are also special digital cables that are designed to carry eight channels of audio at one time. The most common are *optical*, also known as "lightpipe" and sometimes as ADAT format, and TDIF (Tascam Digital Interface). These cables will only work when plugged into compatible devices. Figure 6.16 shows some digital cable connectors.

6.16: Digital cable connectors

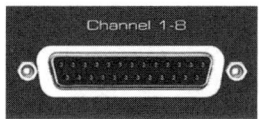

MIDI CABLES

MIDI cables don't carry audio; instead, they send data between instruments and other compatible devices. This data lets you control one device from another—for example, play one keyboard and have it produce sound on another. MIDI actually travels down the same kind of three-conductor wire as stereo audio. Figure 6.17 shows a MIDI cable. We'll explore MIDI in more depth in Chapter 11.

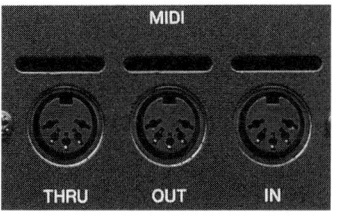

6.17: MIDI connections

DATA CABLES

With computers becoming so important to music, you might also need data cables for your sound reinforcement system. USB cables are used to connect keyboard controllers and audio interfaces to computers. FireWire cables are also used for audio interfaces, as well as connecting computers to hard drives.

Today, many kinds of digital audio equipment, including keyboards and other electronic instruments, effects processors, and preamps, may have data connections that let them interact with computers.

Thus connected, a computer equipped with the right software can be used to play prerecorded music, work as an audio recorder, remotely control devices like mixers, provide audio effects, and even act as a musical instrument that's playable with a keyboard, guitar, drum, or wind controller.

Speaker Connectors

Speakers can connect to amplifiers in a number of different ways. As we saw in Chapter 3, speakers have two terminals: positive and negative. There's no ground.

You can connect to these terminals with raw wire or special lugs, but it's more common to plug into some sort of jack, which is hardwired to the speaker by the manufacturer. Common connectors include mono 1/4", banana plugs, and locking Speakon connectors. Figure 6.18 shows a number of speaker connection jacks and plugs.

6.18: Speaker connectors (from left): 1/4"; banana plugs, Speakon

Wireless Connections

Wireless systems can be used to send a signal from an instrument or microphone to a mixer or amplifier. They can also be used to send a signal from a mixer to a headset worn by a musician on stage for monitoring purposes. Think of a wireless system as an invisible cable that runs between two devices:

1. The transmitter, which sends signal from a source
2. The receiver, which takes the signal and brings it to its final destination

WIRELESS FOR MICROPHONES AND INSTRUMENTS

When you're using a wireless system to send a mic or electric instrument's sound to the mixer or amp, the source (mic or instrument) plugs into the transmitter, which sends a special signal to the receiver. The receiver converts this signal and sends it to its audio outputs, which plug into the mixer.

Wireless mics can either plug into a belt-pack type of transmitter or have the transmitter built into (or plugged into) a mic's housing—a popular option for handheld vocal mics. Figure 6.19 shows both kinds of wireless mic transmitters.

6.19: A clip mic with a belt pack transmitter (below) and a handheld mic with a built-in transmiter (right)

Either way, you need a receiver, like the one shown in Figure 6.20, to receive the signal and plug into the mixer. These receivers sometimes have built-in preamps, and therefore plug into line-level inputs in the mixer.

6.20: A Shure wireless mic receiver. Note the receiver's two outputs: balanced and unbalanced.

WIRELESS FOR ONSTAGE MONITORING

Wireless systems are also used to send a signal from the mixer to individual musicians on stage. Here, the transmitter takes an output from the mixer—such as an auxiliary send—and transmits it to a receiver worn by the musician. A jack on the receiver accepts earphones or *in-ear monitors*—also known as earbuds.

Aside from allowing musicians to have tailor-made monitor mixes, isolating earbuds like the ones shown in Figure 6.21 actually block out some of the other sounds on stage, which allows musicians to hear themselves without turning up the monitor mix. This can protect your hearing. Read Chapter 8 for more on hearing, audio levels, and safe listening.

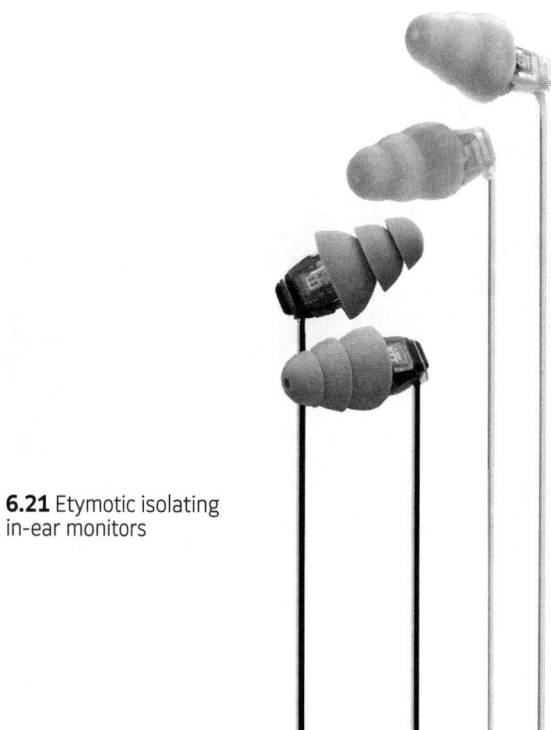

6.21 Etymotic isolating in-ear monitors

ESTABLISHING A WIRELESS CONNECTION

Whether the wireless system is going from stage to mixer, from mixer to stage, or both, the transmitter and receiver must be set to the same channel in order to communicate. By using multiple pairs of transmitters and receivers—each on its own channel—every member of the band can have an independent connection to the mixer.

Consolidating Audio Cables

A stage can get full of cables very quickly. One way to combat this problem is by using a snake, as described earlier in this chapter. When you have a collection of gear close to a mixer, you can also organize all of its various audio connections in one unit, known as a *patch bay*.

STAGE SNAKES

Snakes come in many configurations. The best ones for onstage use have a box on one end and a group of individual cables with connectors on the other.

Typically, the box will house a number of XLR connectors, which can be used to send a mic signal to the mixer. There may also be a few mono or stereo 1/4" jacks, which can be used to send a headphones mix from the mixer to the stage or send the mixer's main outputs to a power amp or powered monitors on stage. These connectors can also be used to route additional audio from the stage to the mixer. The lines going from the stage to the mixer are known as *sends*. The lines that go from the mixer to the stage are known as *returns*.

Figure 6.22 shows an audio snake with eight XLR sends and four 1/4" returns. Audio engineers would therefore refer to this as an "8 x 4" snake.

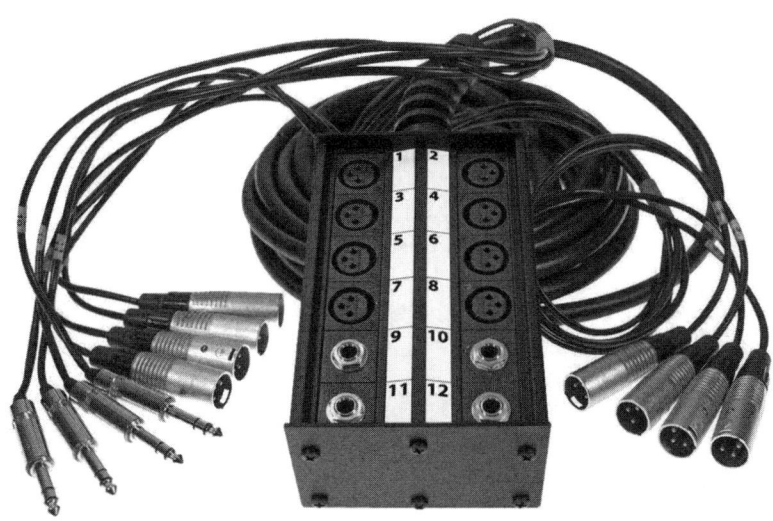

6.22: An 8 x 4 audio snake

PATCH BAYS

A patch bay is a unit with two rows of audio connectors on both the front and the back. These rows are organized into columns (so that each column has a top and bottom row). You plug all your gear into the connections on the back, then use short patch cables to route each piece of gear where you need it to go. Figure 6.23 shows a 1/4" patch bay, but you'll see examples for other types of connections as well.

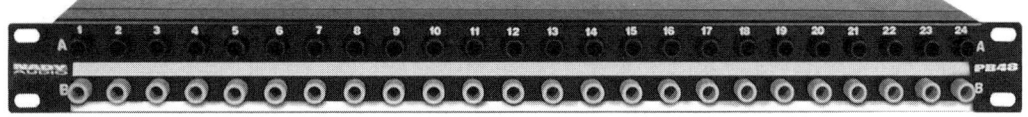

6.23: A 1/4" patch bay

Normalling

Most patch bays use a system called *normalling*. A normalled connection on the back of a patch bay automatically routes any signal coming into a column's top row to its bottom row. Let's say you have a mixer channel's insert send plugged into the top back row of column 1. You can plug the bottom back row of column 1 into the insert return and the signal will flow through that mixer channel as it always does. However, you can reroute the signal (or "break the normal") by plugging a cable into either row on the front of column 1. Figure 6.24 shows how you can use this function to insert a compressor into a channel's signal path.

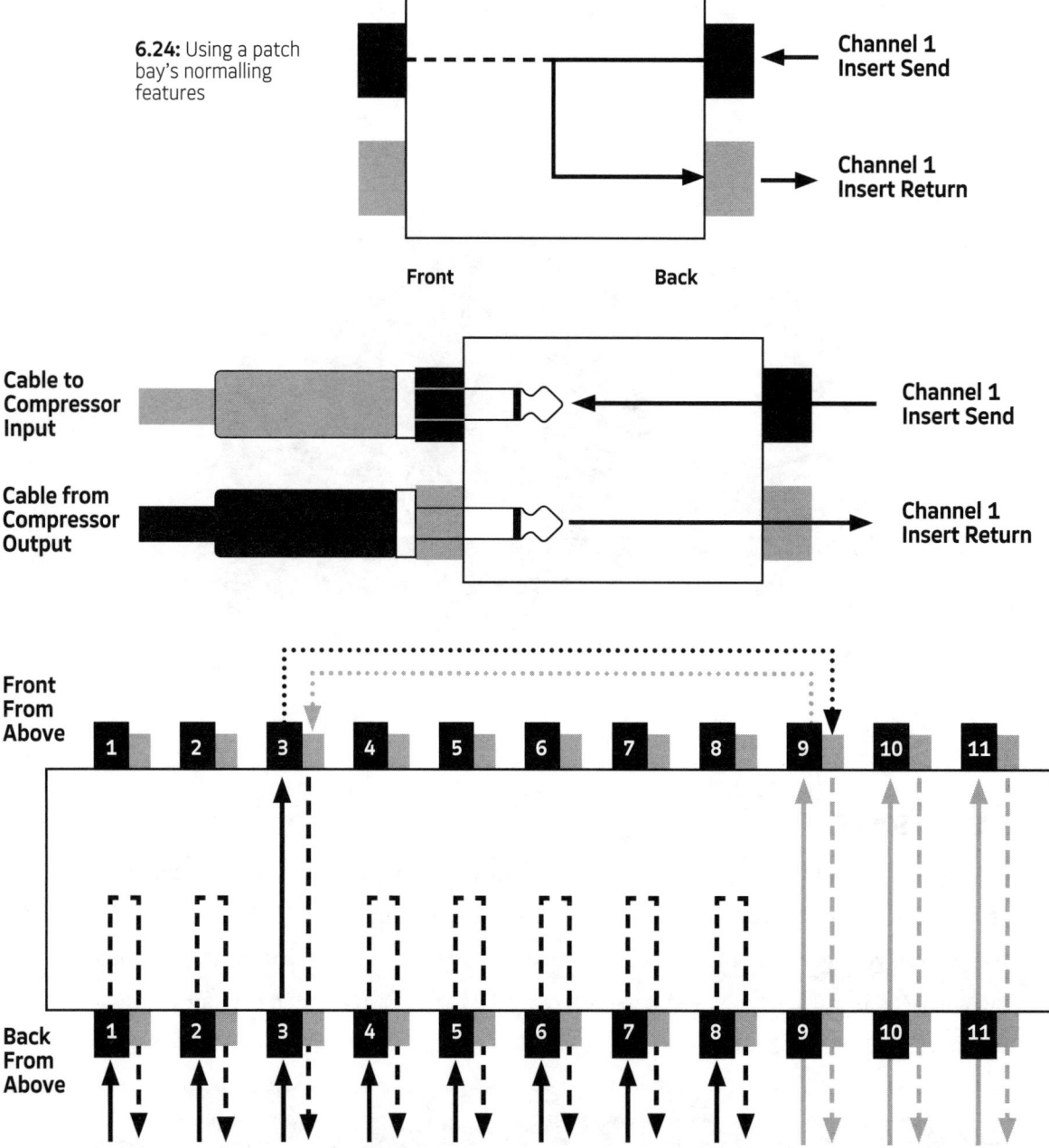

6.24: Using a patch bay's normalling features

Channel 1 Insert Send

Channel 1 Insert Return

Front   Back

Cable to Compressor Input

Cable from Compressor Output

Channel 1 Insert Send

Channel 1 Insert Return

Front From Above

Back From Above

6.25 Using a channel insert with a patch bay. With nothing plugged into the front, signal from the channel's insert send goes straight to the same channel's return, but with patch cords, you can route a signal (column 3 here) to a compressor's input (column 9), then route the compressor's output back to the original channel.

A normalled patch bay is also really useful when you want to access connectors that are hard to reach on a mixing board. For example, a mixer's channel inserts may be in the back of the board, hard to see or get to in a dark performance space. You can plug every insert into the back of a normalled patch bay, running the insert send to the top row and the insert return to the bottom row. If nothing is plugged into the front of any given row, the signal from the send comes back to the return, and that particular channel is unaffected. But when you need to, you can easily patch in a signal processor. Figure 6.25 shows this in action.

SWITCHERS

Switchers can also be useful for routing audio signals. A/B boxes let you route one signal to one of two destinations (or choose which of two signals go to a single destination). These are especially handy for instrumentalists who want to have two instruments plugged into the same amp, but use only one at a time.

An ABY pedal works the same way, but it also lets you have *both* A and B active at the same time.

Plugging Into Electrical Power

Electricity is the lifeblood of every sound reinforcement system. And, unless you're doing an outdoor show in some remote location, electricity is plentiful. So why worry about it?

First of all, electricity can kill you. You should never—NEVER EVER EVER—plug in a piece of gear with a frayed wire or with a broken ground plug. You should also make sure that any circuit you're plugged into can handle the load demanded by all your equipment. That's a question for qualified electricians or the owners of whatever venue you're playing.

If you're playing outdoors, you must take extra precautions to avoid moisture, which can cause electrical shocks. I've had a few of them. They're not fun. But at least I survived. Not everyone is so lucky.

TYPES OF ELECTRICAL CONNECTIONS

Audio devices generally use one of three types of audio connectors: three-conductor AC connectors, which have plugs for hot, cold, and ground; two-conductor AC (alternating current) cords with plugs for hot and cold/ground; and DC (direct current) converters, which plug into an AC outlet and send a low-voltage signal to an audio device.

AC Connectors

AC connectors are pretty universal. In the U.S., every AC device works at 120 volts at 60 cycles per second (this frequency is important for audio).

While AC cables were once hardwired to equipment, most equipment made over the last 20-odd years will have a detachable power cable that follows a standard set by the International Electrical Commission (IEC). IEC cables are usually interchangeable among devices using the same type of connector. Figure 6.26 shows the three-prong IEC C14 connector type that is commonly found on audio gear. Some audio gear may use a two-prong connector, such as the C10.

It should be noted that other countries use different AC current standards than the U.S. For example, an AC connector designed for U.S. users won't work in England, which uses 220-volt power.

6.26: A detatchable grounded power plug

DC Converters

Many devices—most notably small effects, things like wireless transmitters, and even some electronic instruments and compact mixers—use DC power. In order for them to work, their power must be converted from AC with an adapter—one of those big things people sometimes refer to as a "wall wart."

While AC-to-DC adapters are common, they're less interchangeable than AC connectors. Both the voltage and the amperage of the adapter and unit must match, which is one reason why manufacturers specify a particular model for each device they make. DC connectors are bulky, which means they can be a pain in the neck to connect onstage.

Because DC converters are so widely used for stompbox-style effects, manufacturers offer special converters that are designed to connect to multiple pedals in a single chain. Figure 6.27 shows an example.

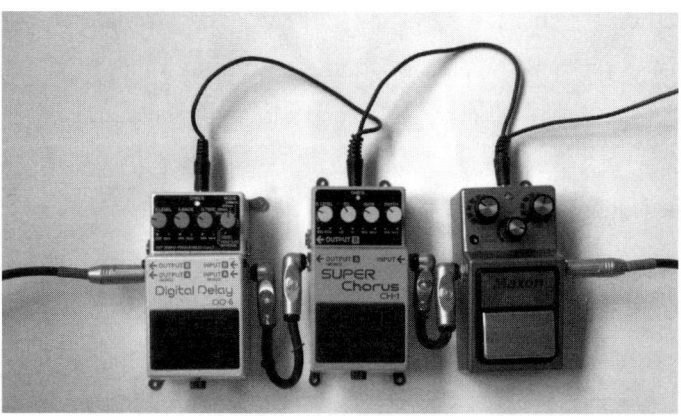

6.27: Pedal effects connected to a single DC adapter

Managing Electrical Connections

If you count up all the connections you'll need for even a small ensemble, you'll soon realize that you'll probably require more electrical outlets than you can reasonably expect to find in a club or auditorium.

Power strips can provide a convenient way to connect multiple pieces of gear. Avoid cheap power strips: they can overheat and actually catch on fire. Better to use a model with a built-in circuit breaker, which will turn everything off if there's a short or other problem, as well as a surge suppressor, which can protect your gear in the event of a power outage, brownout, or other problem with the electricity.

6.28: Power conditioner

POWER CONDITIONERS

Various things can interfere with an electrical signal, and this interference can cause audio problems. Power conditioners can isolate and clean up an electric signal and help your gear sound better. Some, like the unit shown in Figure 6.28, can be rack-mounted—perfect for live sound.

HUM PREVENTION

One of the biggest problems for audio systems is hum caused by interference or by something known as a *ground loop*, which can occur because of faulty wiring or when two pieces of gear that share an audio connection are pulling AC power from different circuits.

There are devices that are designed to eliminate hum (Figure 6.29). One trick to avoid is using a ground lift—one of those gray things people use to plug new equipment into old circuits. The ground is in the circuit for a reason—your safety! Don't remove it!

6.29: Hum eliminator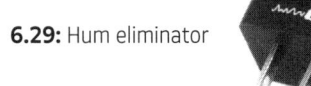

Next Stages

Now that we've gone over the basic components, let's look at a few accessories that can help you organize and set up your gear, before we move on to putting the pieces together.

Chapter 7

ACCESSORIES

So far, we've been looking at the essential pieces of equipment needed for live sound. It's important that you understand how to choose and use such things as mixers, amps, speakers, microphones, effects, and cables—not to mention instruments. But you must also take into account how you're going to get this equipment to and from the show, and how you you're going to set it all up once you get there.

This chapter will take a quick overview of the racks, stands, cases, and other items that can help you get the most efficient use out of your equipment and protect it so that it's ready to serve you show after show.

Racks

One of the cool things about audio equipment designers is that they know when it's time to get together and agree on something. One example is the rack standard, which is pretty universal for all kinds of professional and semi-pro audio gear.

To meet this standard, the front panel of the gear must be 19 inches wide. This width includes the rack *ears* that stick out from each side of the main unit, which fits between the rack's metal *rails*. Each rack ear has two oval holes, which allow you to screw the unit into corresponding holes in the rails.

Audio racks are useful because they allow you to organize and store your gear ahead of time. If you cable everything together in a rack and connect it to a patch bay, you can save a

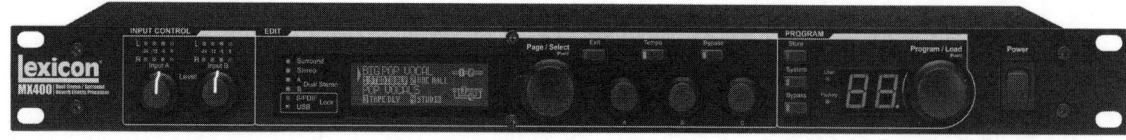

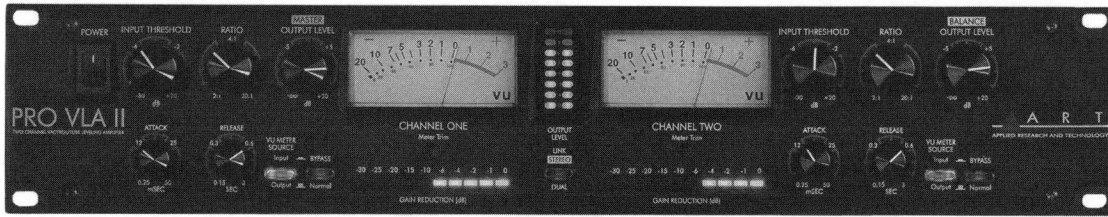

7.1: One-space (top) and two-space (bottom) rack gear

lot of time setting up at the show. Racks are used to hold everything from mixers to effects, power amps, audio recorders, electronic instruments, and even computers.

Rack units also have standard heights: A single rack space (1U in engineer-speak) is 1.75 inches high, so a "one-space" unit (Figure 7.1) will be just under that height, so that it can slot in and out of the rack easily, even when there's gear above and below it. Most audio gear is either one, two, or three rackspaces high, with one exception: rack-mountable mixers can use significantly more spaces.

STANDARD RACK CASES

Standard racks have rails in the front. They can be any height, but usually range from two to 20 spaces, at least for live sound (anything taller would be a pain in the neck to transport). Racks designed for live audio are sturdy—part of their job is to act as a protective case for your stuff. They usually have removable covers, front and back. Figure 7.2 shows a standard rack case.

7.2: A typical audio rack designed for live sound

MIXER RACKS

A standard rack with enough open spaces can mount a portable mixer, but because the controls will be facing forward, they don't put the mixer in an ideal position. You can get racks with additional spaces at the top, specifically designed to accommodate a mixer. Some even allow the mixer to tilt forward for easier access (Figure 7.3).

7.3: A mixer rack with a tilt to make the controls easier to reach

RACK BAGS

Rack bags are portable racks housed in easy-to-carry gigbags (Figure 7.4). These are great for musicians who want to carry personal rack equipment, and are also useful for items you may not want to keep in a larger sound-reinforcement rack (such as outboard compressors or EQs), or items that you might want to be able to take to gigs where your larger P.A. isn't needed.

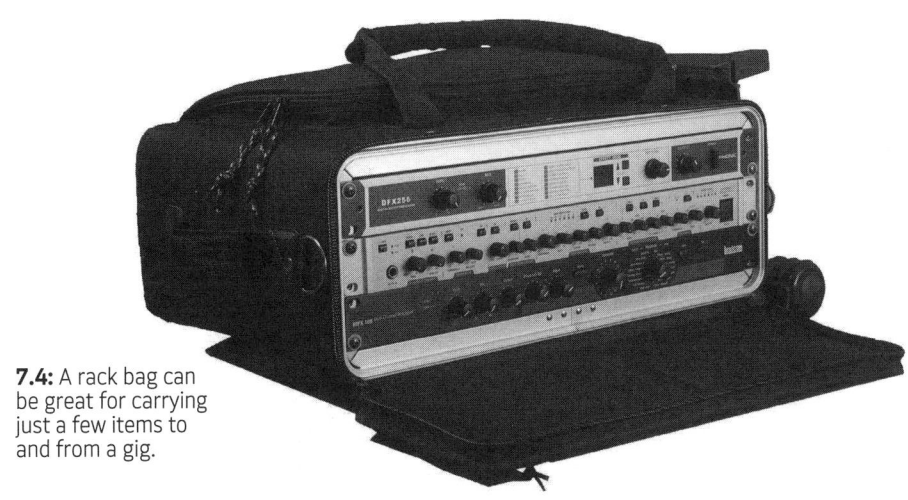

7.4: A rack bag can be great for carrying just a few items to and from a gig.

ADAPTERS, SHELVES, AND DRAWERS

Not every piece of audio gear is designed to be rack-mounted—at least not right out of the box. Fortunately, there are a number of adapters and shelves you can use to rack up this stuff too.

Mixers often require an optional adapter like the one shown in Figure 7.5, which provides a stable base for the mixer and rack ears to hold it in place.

7.5: An adapter designed for a Yamaha mixer

Some units are designed so that two can fit in a single rack space—as long as you have the right kind of adapter or mounting hardware. This is pretty common with things like wireless transmitters. A rack shelf, like the one in Figure 7.6, works well. You can use screws to secure the item to the shelf. Rack shelves come in various heights, and there are even models that slide out. These are especially useful because they can double as a worktable.

7.6: A rack shelf can be useful for mounting non-rack gear.

Drawers are another useful addition to a rack (Figure 7.7). You can use them to hold tools, cables, or adapters. Some professionals also use them to hold stompbox effects, which are then controlled remotely using a switching system.

7.7: A 3U rack drawer

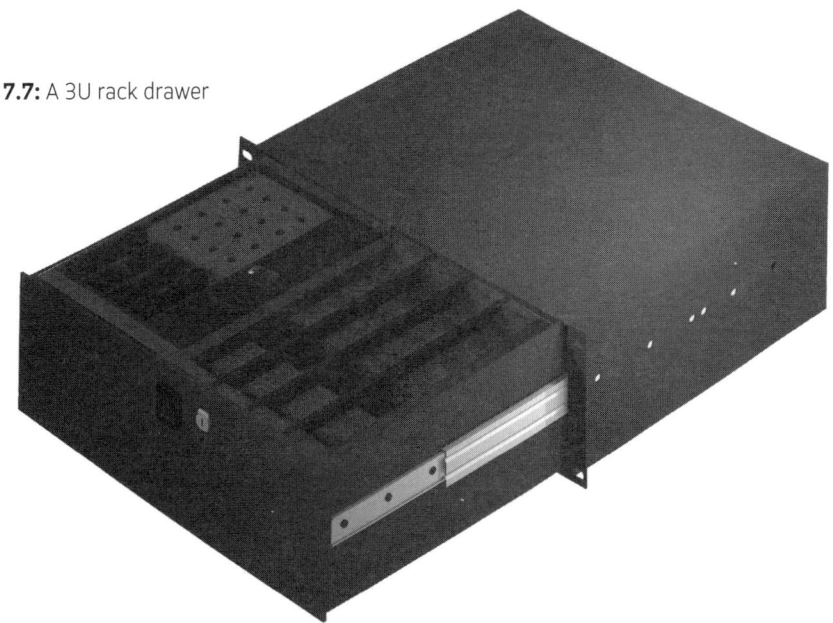

COMPUTER RACKS

Computers are becoming more common on live stages, and you can actually buy rack-mountable models that are designed for audio. However, most computers are not rack-mountable. There are special adapters for desktop and laptop computers. The laptop rack shown in Figure 7.8 is one elegant example.

7.8: A rack device designed to carry a laptop to and from a gig, and provide a home for it onstage

VENTILATION AND POWER

Audio gear can run hot, and it can require a lot of power. Racks are designed to allow air to flow through and ventilate the gear they contain, but you may need to add some additional rack-mountable fans to help things along (Figure 7.9).

7.9: This fan fits in a pair of rack spaces.

In Chapter 6, we looked at power conditioners that deliver clean electrical power to audio gear. Some of these can be rack-mounted. Especially handy are models that provide a front-panel light (Figure 7.10).

7.10: Furman's rack-mounted power strips with front panel lights are useful additions to a live rig.

Stands

A sloppy stage doesn't just look bad; it can also get in the way of your performance. Stands for instruments, amps, speakers, and accessories can make a big difference. Here's a quick overview of useful stands.

KEYBOARD STANDS

Keyboard stands should allow you to adjust your instrument for a comfortable playing position, whether sitting or standing. A simple folding stand like the one shown in Figure 7.11 should do the job if you have only one keyboard.

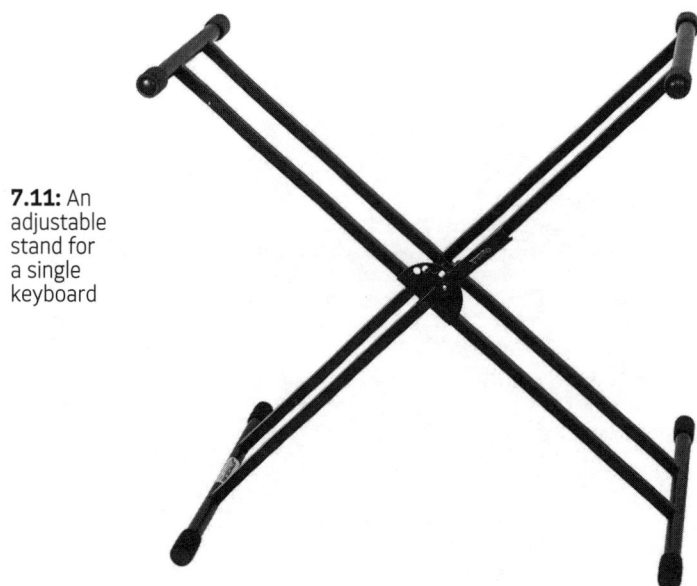

7.11: An adjustable stand for a single keyboard

If you have multiple keyboards, you might want to try a tiered stand, like the one shown in Figure 7.12. These give you easy access to two or three keyboards.

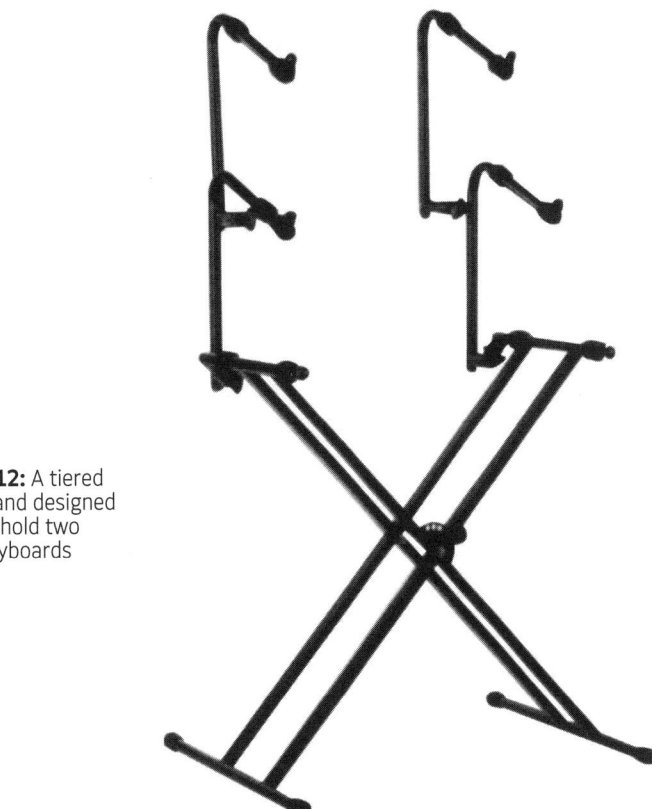

7.12: A tiered stand designed to hold two keyboards

AMP AND SPEAKER STANDS

Technically, stage amplifiers and speaker cabinets don't need to be on stands. They're designed to sit on the floor. But on stage, raising an amp up can help you—and the audience—hear better.

P.A. Speaker Stands

Telescoping stands for PA speakers (Figure 7.13) can really make a difference to your live sound. Instead of having the speaker output blocked by bodies in the crowd or on stage, you can raise it up to where people can hear more clearly. Stands also make it easier to move the speaker to the best location for sound and feedback prevention.

Amp Stands

It's pretty common for guitarists to place their amps on the floor, but this can present problems on stage. If the sound is directed at the player's legs, he or she is going to want to turn up. So the guitar becomes too loud for the audience, which is getting the sound of the amp from stage height. Soon

7.13: An adjustable speaker stand

everyone starts turning up, and the live mix becomes a mess.

Raising the guitar amp and pointing it at the player with a stand like the one shown in Figure 7.14 can help solve this problem. Some feel that the speaker cabinet actually sounds better this way as well, because it's not resonating on the stage.

7.14: Guitar amp stand

An amplifier case, such as the one shown in Figure 7.15, can also double nicely as an amp stand. It protects the amp in transit and provides it a throne during the show. If you do this, make sure the case is designed to accommodate the weight of your amp!

7.15 Using a case as an amp stand

Drum Platforms and Rugs

A drum riser not only allows the audience to see the drummer; it also helps the sound project better. If you're playing a small venue, you probably won't have a drum riser at your disposal. You can actually create a small riser yourself out of plywood pallets, but even that can be a hassle to carry.

Even if you don't have a riser, a drummer should always bring a rug that can serve as a floor for the drum kit. This will keep the kit from sliding on the stage.

MIC STAND ACCESSORIES

We discussed microphone stands in Chapter 4. Just to refresh, you'll find straight stands, boom stands (with an adjustable extension), and devices like goosenecks to make it easier to position the mics. Always make sure that you have the right clips for every mic and every stand before you go to the gig.

A mic stand can also be used to mount other pieces of gear. Some small, light speakers can actually be mounted on top of a mic stand. Perhaps more useful, however, are attachments that let you use the mic stand to hold sheet music (Figure 7.16) or even an iPad (Figure 7.17).

7.16: A music stand that mounts on a mic stand

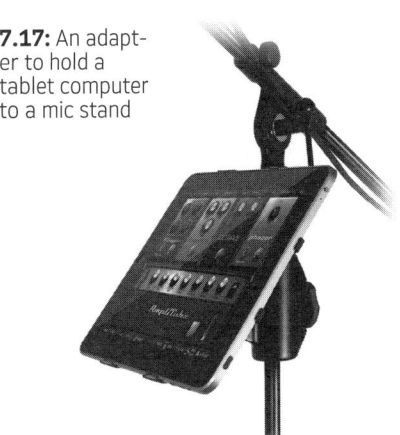

7.17: An adapter to hold a tablet computer to a mic stand

You'll also find such things as holders for guitar picks, instruments, and even cup-holders—a good way to keep water within reach while preventing it from getting knocked over on stage!

INSTRUMENT STANDS

It's important to have a safe place to put down every instrument you bring to a show, and have it ready and in tune when you want to pick it up. Stands are available for every family of instrument. For live performance, look for a folding stand that comes in a self-contained

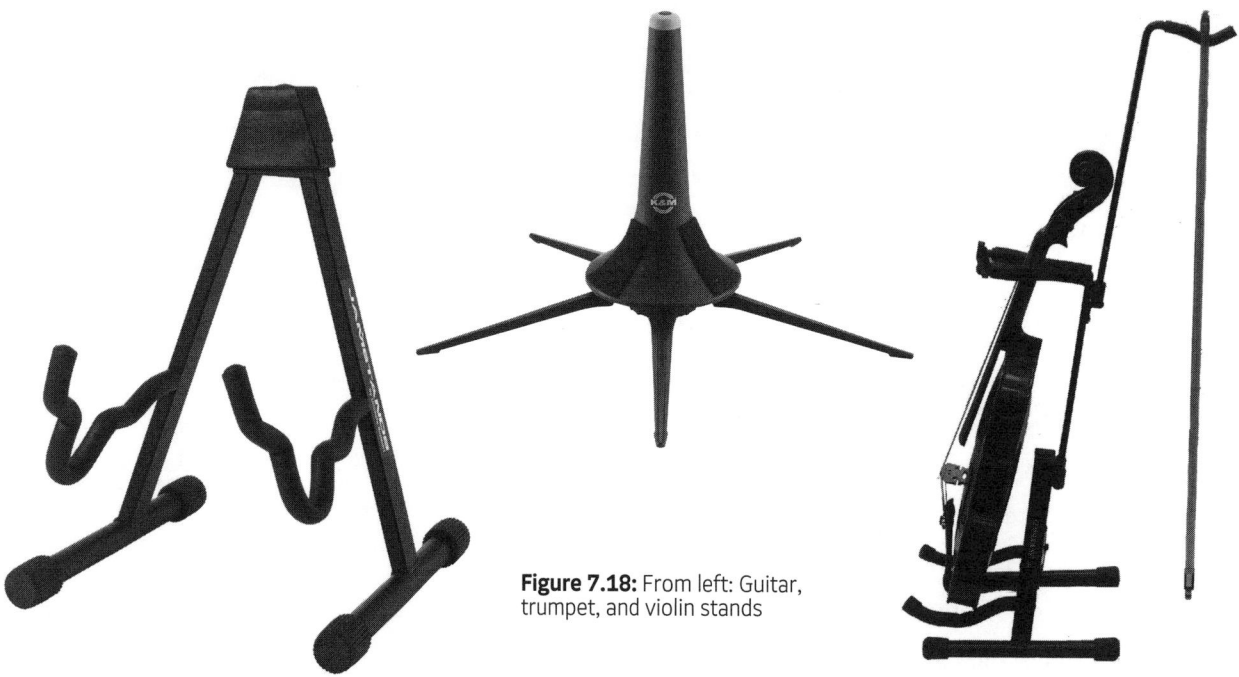

Figure 7.18: From left: Guitar, trumpet, and violin stands

unit (as opposed to those with several parts, which can get lost). Figure 7.18 shows stands for guitar/bass, trumpet, and violin.

If you bring several guitars or basses to a show, you might want something like the multiple stand shown in Figure 7.19, which comes in its own case.

Figure 7.19: A multiple guitar stand with case

Cases

If you plan to play out often, or transport your gear in any way that requires packing, then cases should be high on your list of accessories. As we discussed earlier in this chapter, equipment racks serve as cases and help protect your gear.

For instruments and amps, you have three options: Soft cases, also known as gig bags; hard cases; and flight cases.

Gig Bags

Gig bags are the softest kind of case, and they're the easiest to carry. Some are actually very sturdy, and if packed correctly, can offer as much protection as a hard case. Avoid gig bags that have no padding. Your instrument won't last a car ride in one of those.

Figure 7.20 shows a gig bag designed for acoustic guitar. With its adjustable straps, it's ideal if you're going to a gig on foot or by public transportation.

7.20: A Gator acoustic guitar gig bag

7.21: A set of drum bags by Mapex

Gig bags are available for everything from keyboards to drums and even amplifiers. Figure 7.21 shows a set of bags designed to protect the components of a drum kit.

Hard Cases

Hard cases offer strong protection, but can be heavier and bulkier than gig bags. The best hard cases are those that are designed specifically for your instrument. Not all hard cases are actually hard, however. Cheap cases with cardboard or light chipboard construction can be less protective than a basic gig bag. Look for molded plastic (7.22) or sturdy plywood covered with Tolex. The interior should be plush and padded to hold the instrument in place and prevent scratches.

7.22: A molded plastic saxophone case

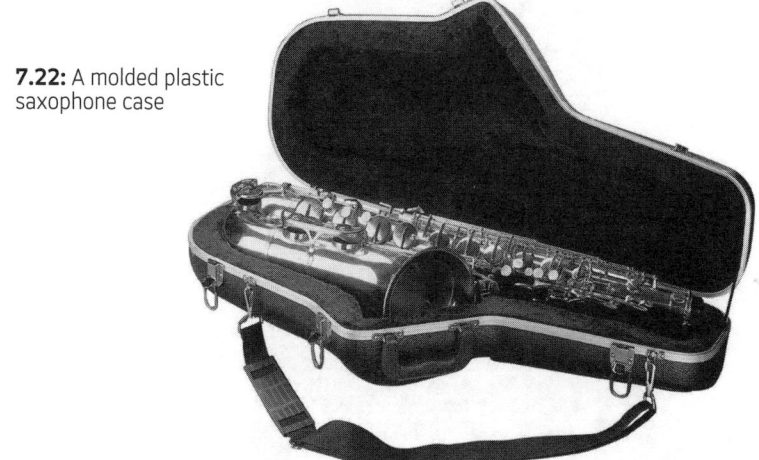

ATA Flight Cases

Cases that meet a specific standard for transport on airlines are known as *flight cases*. These are heavy and expensive, but offer the best protection for your stuff. Unless you're about to go on tour, an ATA case that costs hundreds of dollars is probably overkill.

Pedalboards

Pedalboards allow you to mount and organize floor effects. They're commonly used by guitarists and bassists, but they work for any instrument—or even singers that use effects. You can build your own pedalboard, but there are also commercially available boards (Figure 7.23) in various sizes that let you mount effects to a Velcro surface.

7.23: An SKB powered pedalboard

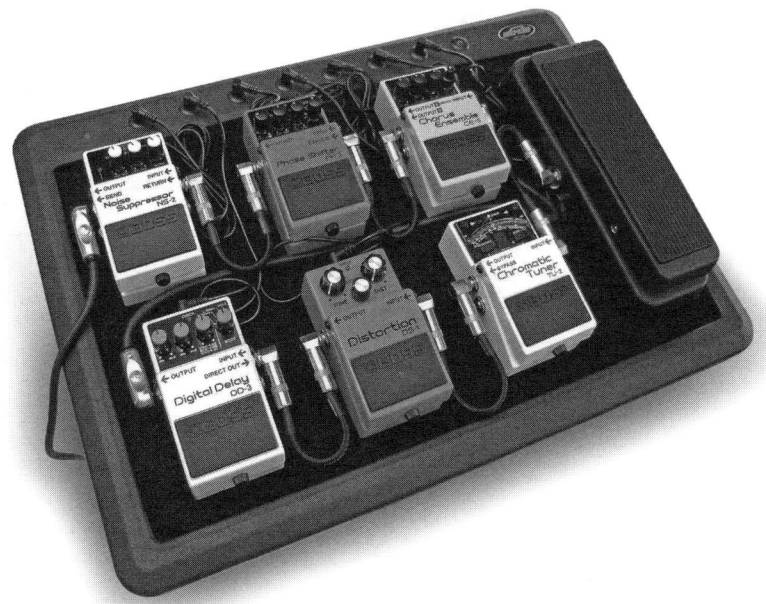

Tools and Tidbits

In addition to the racks, stands, and accessories mentioned above, you should pack a toolkit with the following items:

■ Flat and Phillips screwdrivers in small and medium sizes
■ A pair of needle-nose pliers
■ Wire cutters
■ A flashlight with working batteries
■ Extra 9V and AA batteries for gear
■ A cable tester
■ Sharpies in at least two colors

■ Light masking tape that can be used to label mixer channels
■ Cable ties to help store audio cables
■ A notebook
■ Loose paper for writing set lists (this is a great way to recycle printer paper that's been used on one side)
■ Long and short extension cords
■ At least one spare power cord for audio gear
■ Extra strings, picks, reeds, parts like amp tubes, or anything else that's relevant
■ Cell phone charger

Tuners

Tuners might deserve their own chapter. Certainly, getting in tune is key to putting on a good show. With so many different kinds of tuners available, there's really no excuse for not getting the whole group in tune. Figures 7.24-7.26 show a number of options.

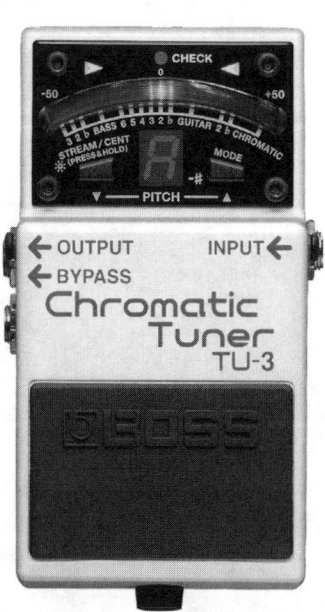

7.24: This Boss pedal tuner is ideal for mounting on a pedalboard.

7.25: A clip-on tuner is great for acoustic instruments because it can be used offstage and on and works even in a noisy environment.

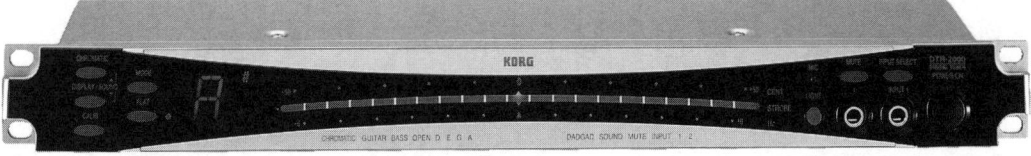

7.26: Rack-mounted tuners that are visible from across the stage are great for players who move around during a show.

Calibrating

Tuners are usually preset at the factory to a pitch reference of A=440 Hz, which has become a standard for electronic instruments. However, if you're playing a venue with a piano or organ that's already in place, you may need to adjust your tuner to match A on that instrument, so having a tuner with an adjustable reference frequency is very useful.

Muting

Most pedal tuners can act as a mute, so that the signal going from the instrument to the amp or mixer is turned off when the tuner is on. You may not want to use this feature, however; it's nice to be able to check your tuning as you're playing. In that case, you should manually mute the amp when you're doing more than a very quick tuning adjustment.

Acoustical Items

If you play often, you'll soon learn that different rooms can have a major impact on your sound. While you shouldn't expect to be able to control this very much, you may be able to reduce a little of a room's rampant echo with acoustical blankets or even with portable acoustical panels.

One way to tame loud sound sources like drums and amplifiers is with sound barriers such as the clear panels shown in figure 7.27. These aren't cheap, but they can be useful if you're playing out a lot and have volume issues.

7.27: Clear sound barrier

Next Stages

Now that we've gone over the basic gear in a live sound system, let's start putting it all together. The next few chapters will look at strategies for setting up, then tackle specific instruments, before bringing it all together in a mix.

Chapter 8

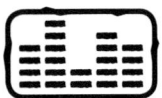

GEARING UP

The first half of this book offers an overview of the key pieces of equipment used in sound reinforcement. We'll continue to talk about gear, especially as it relates to individual instruments, in the next few chapters. But our main focus will be on deciding what kind of sound system you need for the show, setting the system up, and operating the gear so you can get a good sound.

Choosing a System

Let's assume for a second that you have access to a full P.A. with a multichannel mixer, power amp, house speakers, stage monitors, mics, rack effects, and so on. When do you need to use all of it? When will only a few pieces suffice?

Perhaps we should flip this question over: What if you don't have a P.A. system that can handle the venue you're playing, or have a gig at a place that already has a good P.A.? How do you combine what you do have with other available gear?

ASSESSING YOUR NEEDS

The first thing you need to consider is what each member of the ensemble will need. To start, don't even think about whether you have the gear at your disposal or not; just list the things you'll need to get through a show.

Let's say you have a five-piece combo: two guitars, electric bass, drumset, keyboards. Everyone but the drummer sings. One of the guitarists plays both electric and acoustic guitars. Both guitarists use pedalboards for effects. The keyboardist also doubles on saxophone. Here's the minimum that you'll need.

Guitarist 1: Electric guitar, amp, vocal mic, electrical outlets for the amp and pedalboard, two instrument cables, one mic cable.

Guitarist 2: Electric guitar, amp, vocal mic, electrical outlets for the amp and pedalboard, acoustic guitar, D.I. for the acoustic guitar, three instrument cables, one mic cable.

Bassist: Bass, amp (possible D.I. for the bass), vocal mic, electrical outlet for the bass amp, one instrument cable, one mic cable.

Drummer: Six-piece drum kit, sticks, stool.

Keyboardist: Keyboard, stand, amp or stage monitor for the keyboard, electrical outlet for the keyboard (and amp, if used), D.I. inputs for the keyboard, vocal mic (doubles as sax mic), one instrument cable.

Counting Audio Inputs and Electrical Outlets

Let's say this band is playing a relatively small room where neither the drums nor the electric guitar amps need to be miked. You're still going to need a mixer with a minimum of seven inputs: four for the vocal mics, and one each for the acoustic guitar, bass, and keyboard D.I. If you want to run the keyboard in stereo, you need eight inputs.

You'll also need at least six electrical outlets. Each guitarist will need two (one for the amp, one for the pedalboard). The bassist needs one. The keyboardist needs one. If the keyboardist is using an amp, you need a seventh. Of course, none of this accounts for the electrical power needed to run the P.A. If you're bringing it, assume you'll need an outlet for your mixer, amps (or powered monitors), and so on.

The Backline

The lineup of equipment described above—guitar amps, bass amp, drum kit, etc.—is known as a *backline*, so named because it's usually positioned at the back of the stage. If you're in any kind of ensemble that uses amplified instruments and drums, you can think of the backline as the backbone of your sound.

Figure 8.1 shows a typical backline. As you can see, there are amps along with a drum set on a riser.

USING A PRE-EXISTING BACKLINE

Professional musicians who play large venues carry their own backlines on tour, and if you're setting up all your own gear for a performance, you'll need to bring the amps and drums to set up your own backline.

8.1: A backline with guitar amps, a bass amp, and drum set

However, some venues—especially those featuring more than one act on any given day—will often provide a backline that's already set up. They want you to plug into their stuff and play through it because it both saves time and ensures that the equipment is already known to be working and configured for the house mixer. Often, you can find a venue's backline information on its website. If in doubt, it pays to call ahead.

Let's say our five-piece band is playing a venue with a pre-existing backline consisting of two guitar amps, a bass amp, a five-piece drum set, and a 16-channel mixer. By consulting the list we made earlier, we can see that the guitars and bass are covered by onstage amps. The drummer is playing a smaller kit, but can probably live with it. The keyboardist will have to go direct to the board (or bring an amp). The mixer has plenty of inputs for the vocal mics and D.I. for the acoustic guitar.

ADDING TO OR MODIFYING A VENUE'S BACKLINE

When you call a venue to ask about the equipment, it pays to ask if you can bring anything of your own. Drummers can usually bring their own snare and cymbals.

Guitarists can be choosy about amps, and it's not uncommon for them to play through something totally alien when it's part of a new venue's backline. One way around this is to use a preamp (either rack-mounted or as part of a pedalboard) and feed that to an amplifier from the venue's backline.

You can communicate your backline needs as part of your *stage plot* (see the next sections).

SETTING UP A BACKLINE

If you're setting up your own backline, you need to account for several things. Where do you put everything? How do you position the gear? How do you connect the gear to the instruments and—if necessary—P.A.?

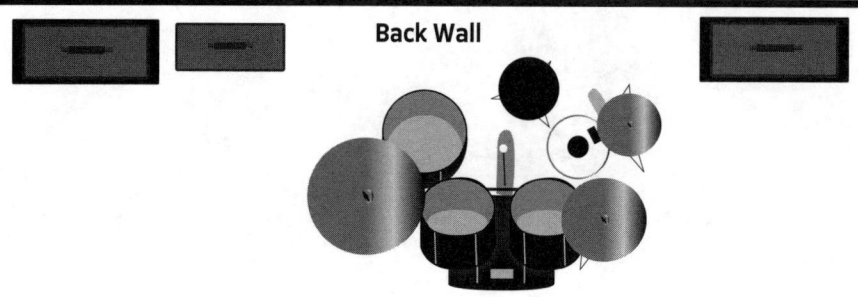

Back Wall

8.2: A simple backline on a small stage. Everything is close to the back wall.

Front of Stage

Back Wall

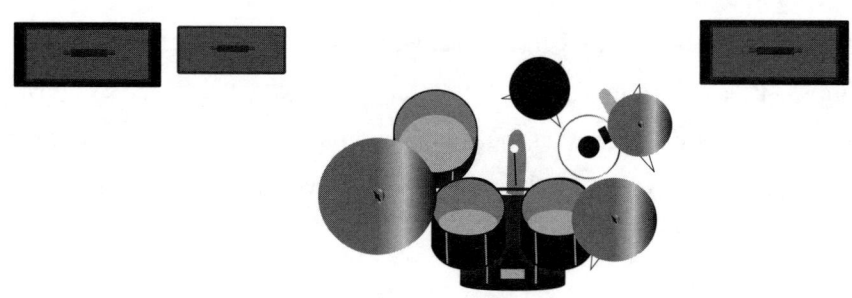

8.3: The same backline on a deeper stage. It's important not to set up too far back from the audience.

Front of Stage

Preexisting Stages

The ideal location for a backline is any uninterrupted space that can accommodate the drums and amps. If you're playing on a shallow stage, the backline's location is pretty obvious: it goes against the back wall (Figure 8.2).

If you're playing a deeper stage, such as a school auditorium, the backline can move away from the back wall. It should be deep enough to give the musicians room, but not so deep that the drummer is too far away from the rest of the players (Figure 8.3).

When you set up, try to account for how the drummer will hear the rest of the band. If

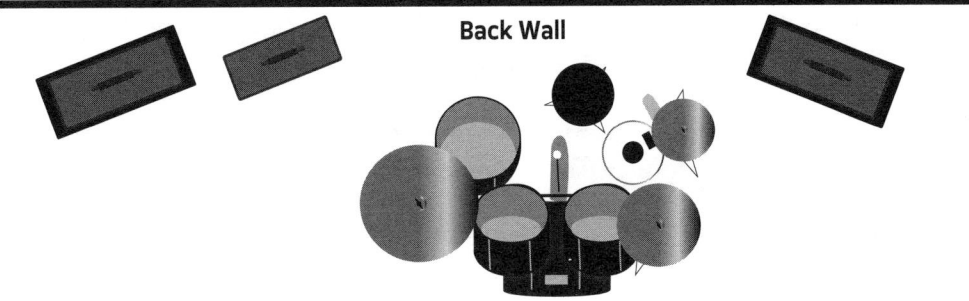

Back Wall

8.4: On a shallow stage, you may need to angle the amps so that the drummer can hear.

Front of Stage

Back Wall

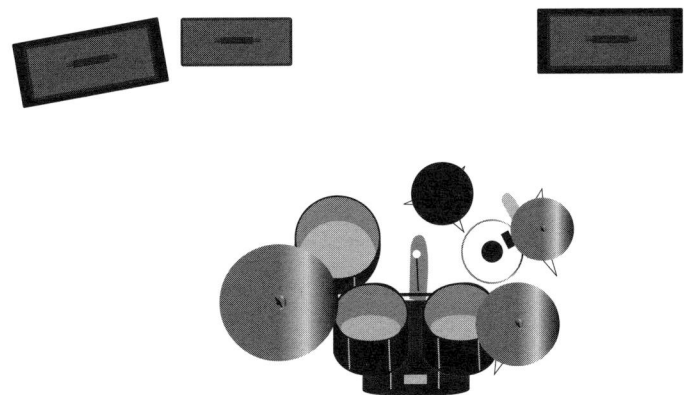

8.5: On a deeper stage, amps can go behind the drummer.

Front of Stage

the drummer doesn't have a monitor mix, the onstage amps will need to be positioned so that he or she can hear them. In Figure 8.4, the guitar and bass amps are angled slightly on the shallow stage so that the drummer can hear.

On a deeper stage, the amps can actually go just behind the drummer, so he or she can hear more easily, as shown in Figure 8.5.

If you're playing a stage at an outdoor venue with a canopy, the backline should be set up so that all of the electronic gear is covered, in case it rains.

If you're in a space without a pre-existing stage, you have more options, but also face more challenges. Not all venues have enough room for the backline to spread out. You might have to cram yourself and your gear into some tight spaces.

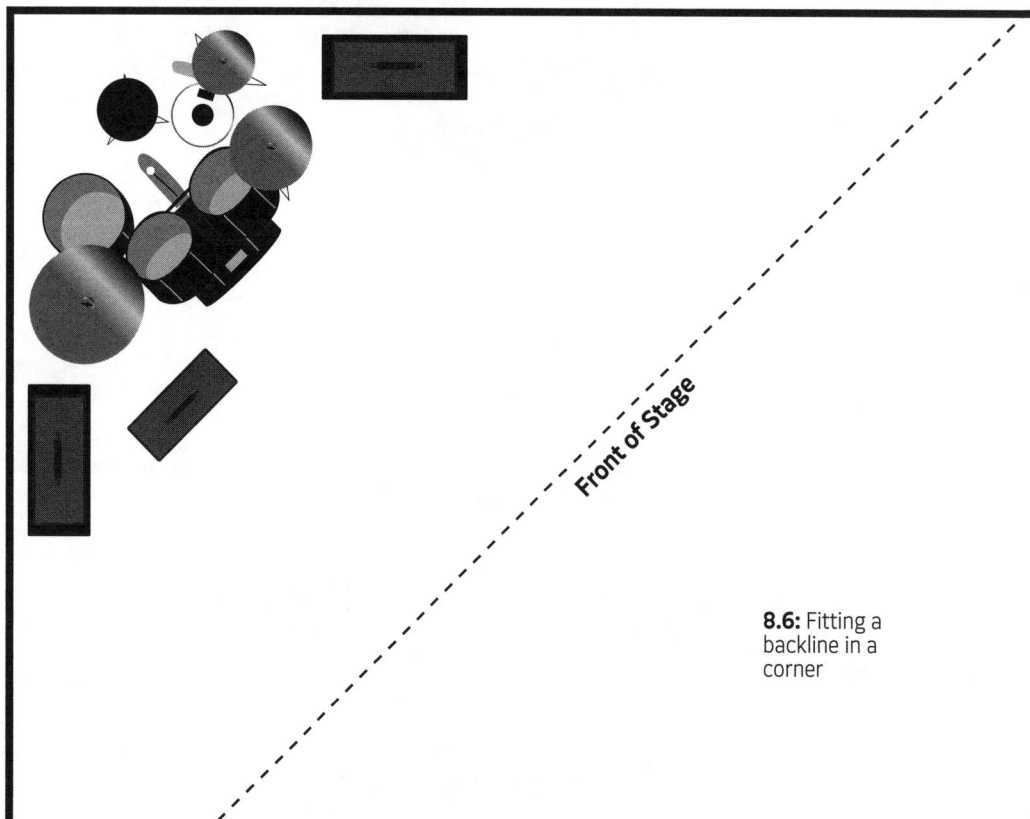

Front of Stage

8.6: Fitting a backline in a corner

Makeshift Stages

Plenty of venues require you to create a "stage" by taking over any part of the room you can adapt into a performance space. Maybe it's the corner of a school gym or a part of a library community room, a friend's living room, patio, or a section of a restaurant.

Unfortunately, these rooms probably have a few obstacles that might get in the way of your ideal backline. You might find yourself shoving the drums into a corner, with the amps on each side. This not only makes it hard to set up, but also makes for a crowded performance space. Figure 8.6 shows a backline in a tight space. Not ideal, but it'll work.

Accounting for Connections

Whether you're setting up on a formal stage or makeshift performance space, you must consider two factors:

     1. How to get electrical power to the backline

     2. How to make audio connections to the gear

The problem of electrical connections is easy to solve with a heavy-duty extension cord and one or two good-quality power strips or power conditioners. Run the extension cord as far away from the audio cables as possible to the back of the backline, and position the power strips so that every piece of gear can plug in without stretching its power cable. (You may need a second extension cord for equipment positioned at the far end of the drum kit.)

The audio connections aren't exactly complicated, but they can take some planning. Audio cables need to be long enough to reach the amps, but shouldn't be so long that they get in the way on a cramped stage. There needs to be room for pedalboards on the floor. Cables should be secured so they don't come out of the amp or pedalboard and don't dangle and

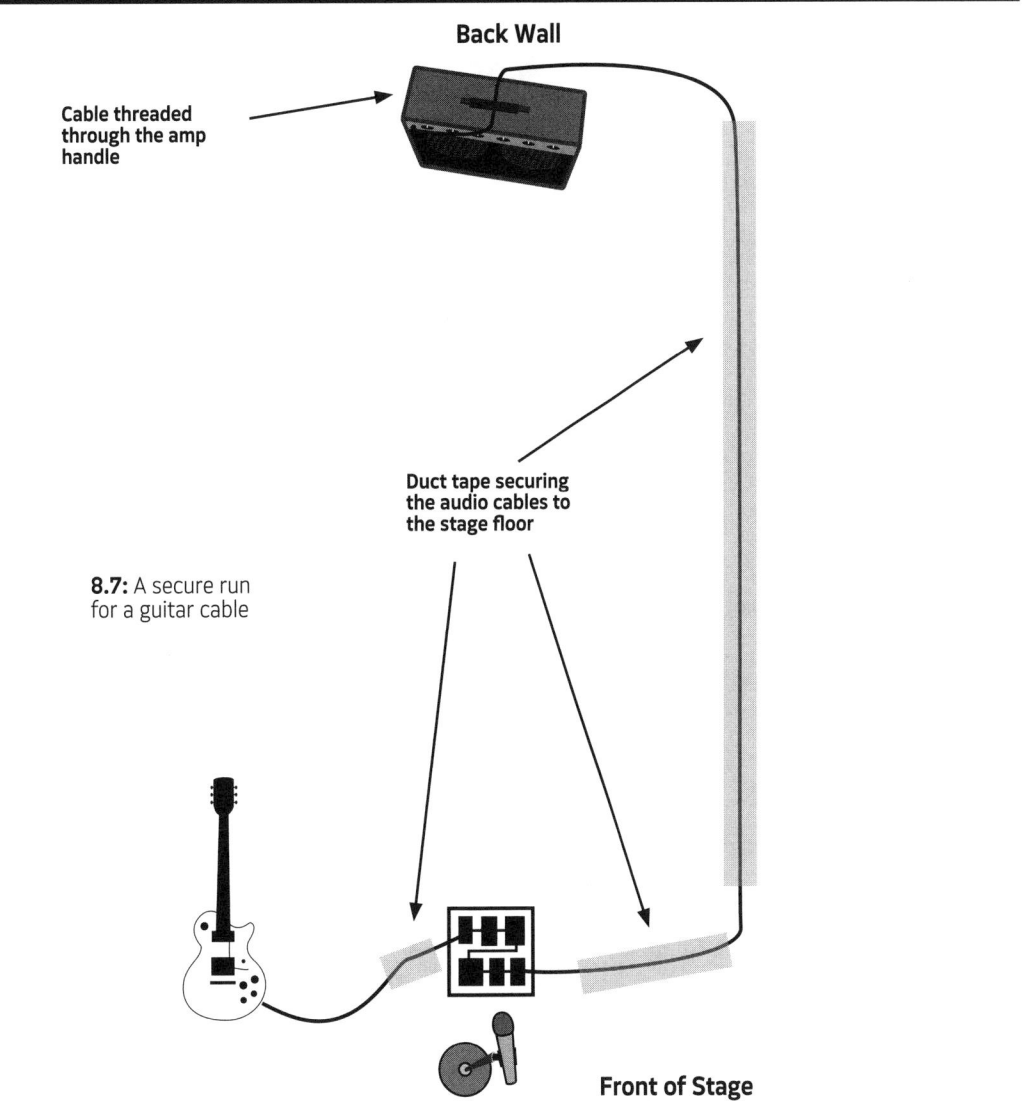

Back Wall

Cable threaded through the amp handle

Duct tape securing the audio cables to the stage floor

8.7: A secure run for a guitar cable

Front of Stage

trip the performers. By the way, these aren't hypothetical situations: Calamities between people and cables happen all the time.

Figure 8.7 shows how to run a cable from the front of the stage to an amp positioned in the backline. Note that the guitarist is using a short cable to the pedalboard, and a longer cable between the pedalboard and amp. The cable going to the amp is taped down to the floor with duct tape and secured at the amp by running through the amp's handle.

We'll look at some other instrument-specific options in Chapters 10 and 11. But for now,

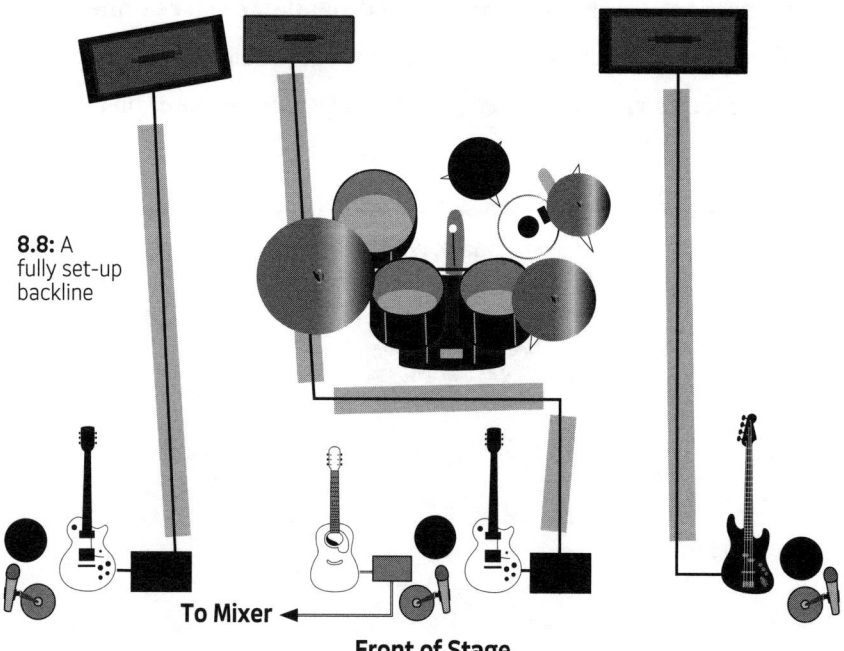

Back Wall

8.8: A fully set-up backline

To Mixer ←

Front of Stage

think big picture: Every instrument is going to need to connect to its amp (or the mixer) securely without causing a tripping hazard on stage. Figure 8.8 shows a backline that's been set up, with cables running to the appropriate spots on stage. Note the addition of mic stands. This diagram could actually make a fancy stage plot (though it would be overkill for that purpose).

Solo, Duo, and "Non-Backline" Lineups

Of course, not all acts use bass, drums, electric guitars, etc. Solo acts and duos may need smaller setups. Acoustic ensembles may require a completely different approach to amplification. That doesn't mean you don't need to account for your inputs and outputs. Here are a few examples of how to assess your onstage sound needs.

YOUR INDIVIDUAL LIST

Every member of the ensemble must be able to identify and communicate what he or she needs to sound good on stage. Look at your entire set: What's your main setup? Will you be changing gear or instruments? If so, when and how? And how will that affect the inputs or amps you need?

If you're a solo performer, then you're the whole ensemble. You still have to make a list of inputs—and it may not be as simple as you think.

Let's say you're a singer-songwriter who plays acoustic guitar. You'll need to amplify your voice and amplify the guitar. So you need at least two inputs.

Your voice is easy: You can sing into one mic. But are you going to stand or sit? If seated, what kind of stool do you need? Do you need a music stand? What about the guitar? It may or may not have a pickup. If it doesn't, you'll need at least one more mic. If it does have a pickup, you'll need a way to plug it into the mixer, which may involve using a direct box or a preamp. Do you have your own preamp? This might require a different kind of mixer input. What about effects? Do you use stompboxes between the guitar and the input? Do they require power? Perhaps you have an acoustic guitar amp. Do you need it? If you do bring it, will it be loud enough to amplify your guitar in-house or serve as a personal monitor for you? Will you be sending a feed from your amp to the house mixer, or will you use a direct box between the guitar and the mixer?

What if you want to bring a second guitar? Do you want to have that connected ahead of time? If so, you'll need a second guitar input. Or perhaps you might unplug the first guitar and replace it with the second from time to time during the set. You can do this without adding inputs, but you may want to let the sound engineer know so he or she can mute the mixer when you do. If the second guitar's output is louder or quieter than the first's, the engineer will also have to adjust the mixer.

DUOS

If you're in a duo, you'll be dealing with many of the same questions as a solo performer, only doubling them. In many cases, each player can bring his or her own amp or small P.A. In others, you'll have to share gear, or explain your needs to the person running a venue's sound system.

ACOUSTIC ENSEMBLES

If you're in an acoustic ensemble, you'll need to assess your group's natural balance before deciding what amplification is required. If the venue is small, you may use the P.A. for minimal enhancement and let the audience hear as much of the acoustic instruments' natural tone as possible. Maybe you need a single mic on the side to use for a singer or when an instrument is soloing. Or maybe one very quiet instrument needs to be amplified.

In a larger venue, you might route each instrument through its own channel (or channels) on the P.A., or you might opt to capture the whole unit with overhead microphones.

MIXING ENSEMBLES

Things can get tricky when you combine acoustic instruments with electric guitars, bass, keyboards, or percussion. I recently witnessed a school performance with an orchestra and pair of electric guitars. Unfortunately, the guitar amps were set up in front of the orchestra, instead of in the backline. They drowned out the violins, violas, cellos, and string basses for the audience, yet because they were facing away from the conductor and orchestra, the players on stage had no idea this was happening.

One way to prevent this kind of scenario is by creating a stage plot.

The Stage Plot

A stage plot is a simple diagram that shows every instrument in its position on stage, and makes a list of the inputs and power requirements needed. It may also include notes (such as whether a player is sitting or standing, or whether one player may move to different instruments during the show). Drawing up a stage plot is one of the best ways to assess your performing needs and to communicate what they are to others.

Stage plots are useful for both the musicians and the people operating the sound. If the musicians are providing their own sound equipment, a stage plot can help them create a packing checklist (see Chapter 14) and can also be used to let the venue owner know how much room and electrical power is needed.

Figure 8.9 shows a basic stage plot. As you can see, it's nothing fancy. But it offers a lot of information.

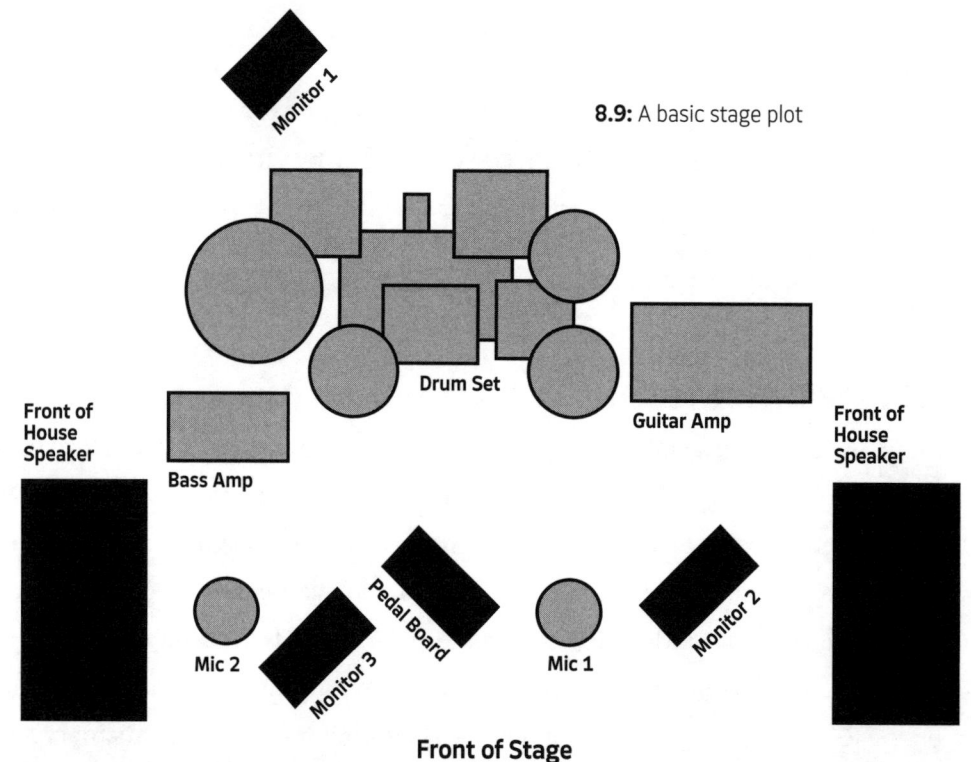

8.9: A basic stage plot

If you're responsible for your own sound, a stage plot can help you plan for a show by illustrating what you need to bring and where to set everything up. And if you're not doing your own sound, a stage plot is even handier: With a stage plot and a well-marked set list, a sound engineer who's never seen you before has a much better chance of making you sound good.

Let's start by plotting out a single performer's stage setup, then expand to show how you might plot out a whole band.

STAGE PLOTS FOR SOLO PERFORMERS

A solo performer might assume that any venue has what he or she needs, but doing so leaves a lot of things to chance, so even here, a stage plot can be pretty useful.

Figure 8.10 shows a stage plot for a singer-songwriter who's working with one pickup-equipped guitar.

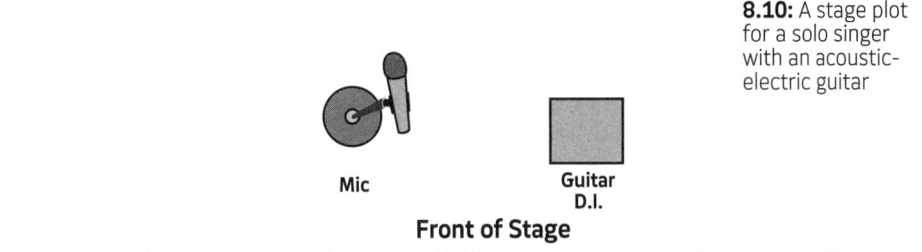

8.10: A stage plot for a solo singer with an acoustic-electric guitar

Mic

Guitar D.I.

Front of Stage

Figure 8.11 shows a slightly more complex setup for the same musician. The majority of the show will be played standing with the pickup-equipped guitar. But several songs will be played seated with a non-pickup guitar, which needs to be miked.

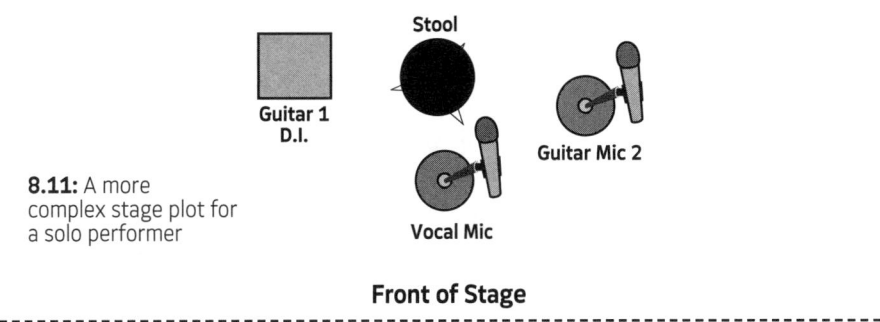

Stool

Guitar 1 D.I.

Guitar Mic 2

8.11: A more complex stage plot for a solo performer

Vocal Mic

Front of Stage

Figure 8.12 shows a stage plot for a solo performer who uses both guitar and keyboards. To save time during the switch from guitar to keyboard, the performer has asked for two vocal mics: one center stage and one by the keyboard. You'll also notice that the keyboard has stereo outputs and requires a 120-volt outlet.

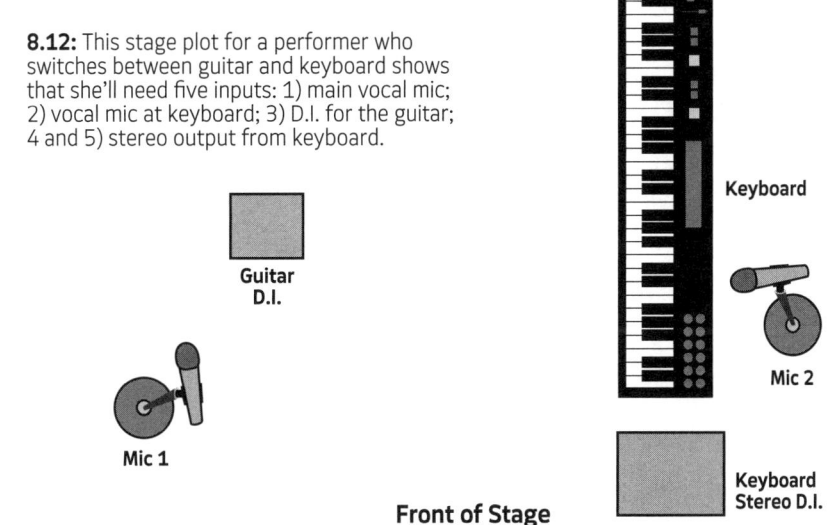

8.12: This stage plot for a performer who switches between guitar and keyboard shows that she'll need five inputs: 1) main vocal mic; 2) vocal mic at keyboard; 3) D.I. for the guitar; 4 and 5) stereo output from keyboard.

Guitar D.I.

Keyboard

Mic 2

Mic 1

Keyboard Stereo D.I.

Front of Stage

The same thinking can apply when you're setting up your own system. Let's say you're playing a venue that has no built-in sound system. The scenario shown in Figure 8.10 could be handled by a small acoustic guitar amp that has a channel for a vocal microphone, like the one shown in Figure 8.13. A small all-in-one P.A. would also work pretty well. The setup in 8.11 would require a small all-in-one P.A. or an acoustic guitar amp with two microphone channels. Scenario 8.12 would require either two amps (one for the acoustic guitar and the other for the keyboard, with one or both providing a channel for vocal mics) or a small P.A. with at least five channels. A small mixer and one or two powered monitors could also get the job done.

8.13: An acoustic guitar amp with a microphone input may be all you need for solo performance.

Unless you've got a warehouse full of gear, you probably don't have all those options at your disposal. You'd have to choose between, say, an acoustic guitar amp and a portable P.A. system. Knowing what you'll be using most often will help you make the right choice, while understanding how every option might work can help you adapt when you don't have the exact gear needed.

STAGE PLOTS FOR ENSEMBLES

Now let's look at a larger ensemble. Figure 8.14 shows a photograph of an actual stage plot used by a touring rock band called the Afters. The picture was taken at the soundcheck for an outdoor concert. It diagrams everything: the backline, the mics, instruments, and power

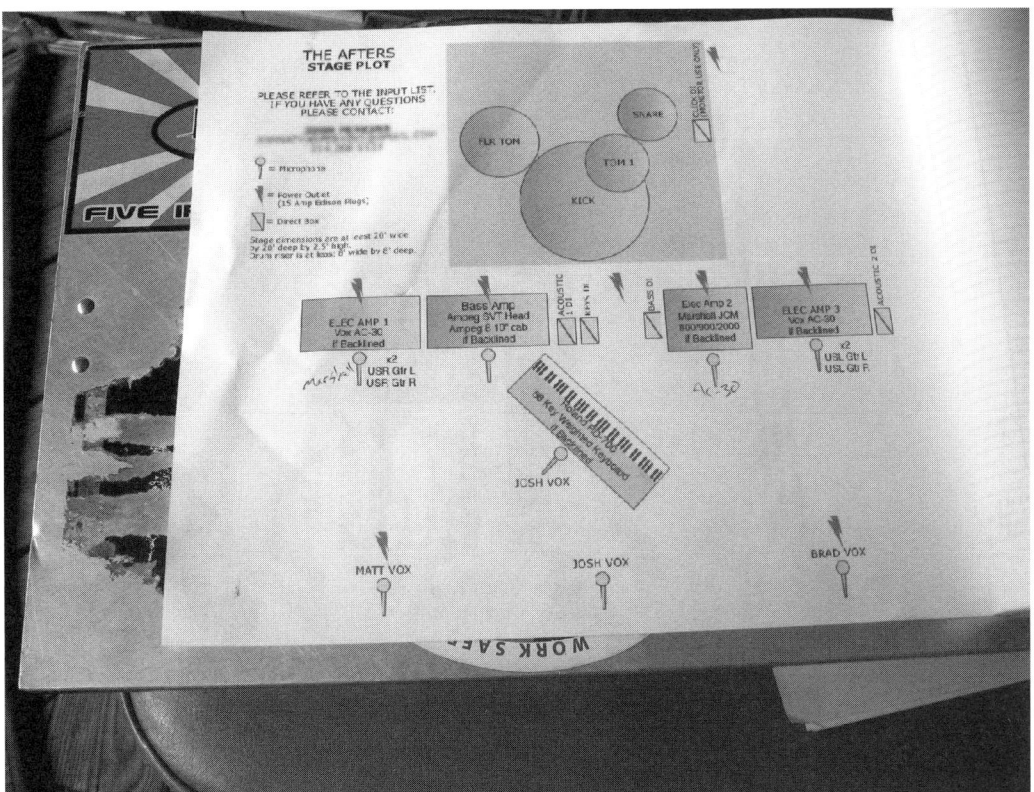

8.14: A detailed stage plot used by rock band the Afters for an outdoor performance

requirements. It even shows how one of the singers, Josh, needs two mics: one at center stage, the other at the keyboard positioned a little further back (note that it also shows how Josh wants the keyboard angled to face Matt). This stage plot also identifies the musicians by name and gives contact information for the band's road manager.

A diagram providing this much detail will help you (or a venue's sound engineer) set up the entire sound reinforcement system to best serve both you and the audience.

Next Stages

With a stage plot, you can start to plan how to set up your P.A. system. Remember, the requirements will change depending on the number of players, the instrumentation, and the size of the venue.

Chapter 9

SETTING UP
YOUR SOUND SYSTEM

Setting up a P.A. system can be a pretty complex undertaking, so it pays to be organized—and to understand how much P.A. you actually need for the situation at hand. For the moment, let's assume you're facing one of four scenarios:

1. A small gig that you can cover with an all-in-one P.A., small portable system, or even a single amplifier.

2. An ensemble gig at a small venue that will require a P.A. of moderate power, with only vocals and acoustic instruments running though the P.A.

3. An ensemble gig at a larger venue that will require a P.A. of moderate power, with drums, vocals, acoustic instruments, and possibly amplifiers being miked through the P.A.

4. A larger venue where every instrument—including your stage amps—will be fed through the house P.A. system, operated by one or more professional audio engineers.

In scenario 1, the performer will almost always be serving as his or her own sound engineer from the stage. In scenarios 2 and 3, the mixing console might be positioned on or off the stage, and might be operated part-time by a band member or full-time by someone in your crew or someone on the venue's staff. In scenario 4, the sound engineer will be offstage operating the mixer throughout the show.

Mics, Speakers, and Feedback

Before we get into specific setups, let's look at one problem that plagues live performers: feedback. You'll encounter two kinds of feedback: *microphonic* and *resonant*.

MICROPHONIC FEEDBACK

Microphonic feedback is that high-pitched squeal that happens when a microphone picks up its own sound as it comes through a speaker system. Audiences hate it, and it can actually damage your ears. During soundcheck, professional live sound engineers do a procedure called "ringing out the room" to find out which frequencies are most prone to feedback, and you can use automatic feedback eliminators to help tame the problem. But the best thing you can do is avoid pointing your speakers at your mics.

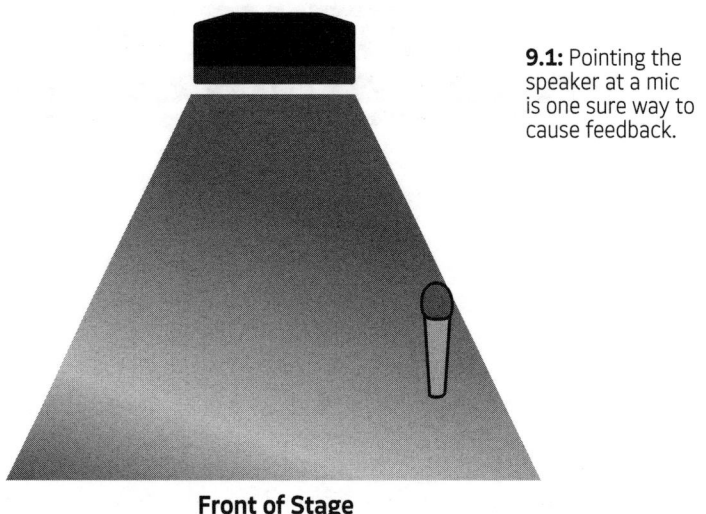

9.1: Pointing the speaker at a mic is one sure way to cause feedback.

Front of Stage

Figure 9.1 shows a mic/speaker relationship that's likely to cause feedback. The house speaker isn't even pointing directly at the mic, but it is putting out enough sound in the mic's direction to cause a problem. Figure 9.2 shows the same speaker, moved away from the mic, so that the mic's pickup pattern isn't capturing any of the speaker's output. Because the speaker has been moved, it will be harder for the singer to hear. A stage monitor behind the mic, pointing up at the singer from the floor, will solve that problem. This setup isn't "feedback proof" but it will help, as long as the volume levels don't get too loud.

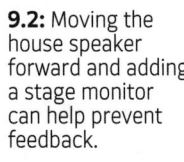
9.2: Moving the house speaker forward and adding a stage monitor can help prevent feedback.

Stage Monitor

Front of Stage

RESONANT FEEDBACK

Resonant feedback occurs when an instrument starts to resonate at a particular frequency and creates a sustaining note. This can plague acoustic stringed instruments—if you think about it, the top of a guitar or cello is like a speaker and a microphone rolled into one. If your instrument is resonating only when you're holding it directly in front of a speaker, simply moving or turning away from the speaker may solve the problem. You might also adjust the EQ to tame the offending frequency, move the instrument to a location that is less resonant at that frequency, or use something to dampen the instrument and prevent the problem (this will actually affect the tone, so try to avoid it if you can).

Small Venues

Small venues may not require much in the way of a P.A. system. Your goal will be to enhance the natural sound of your instrument(s) and voice—unless your music is purely electronic and therefore requires amplification, no matter what. We'll start with a solo setup and build from there.

SOLO PERFORMERS IN SMALL VENUES

Solo setups are easy in the sense that there's not much to connect, but if you're playing, singing, *and* setting up your stuff, you must consider three things:

1. Where will the system best serve the audience?
2. How well can I hear myself?
3. How easily can I access the controls?

If you're a solo keyboard or acoustic guitar player who also sings, you could use an acoustic guitar amp that has both instrument and microphone inputs with separate level controls.

If you place the amp behind your playing position as you face the audience, you'll be able to hear yourself, but you must be careful that the amp's speakers don't point at the mic, which can cause feedback. If you put the amp in front of you, you won't hear your amplified sound as much. In a quiet space, this might not be a problem, since your instrument and voice will be audible acoustically.

Amps vs. Small P.A. Systems

Some instrument amps act like mini P.A. systems, but are they always the best option? The advantage of using an amplifier is that its controls and overall sound will be tailored to your instrument, so it can be easier to get the instrument sounding great without adding gear. On the other hand, a combo amp, in which the controls and speaker are together, offers less flexibility when you're setting up.

A small all-in-one P.A., or system with a small mixer and powered monitors, allows you to place the controls near you and move the speakers to best advantage. If the system has stereo speakers, you can face one toward the audience and use the other as a monitor (Figure 9.4).

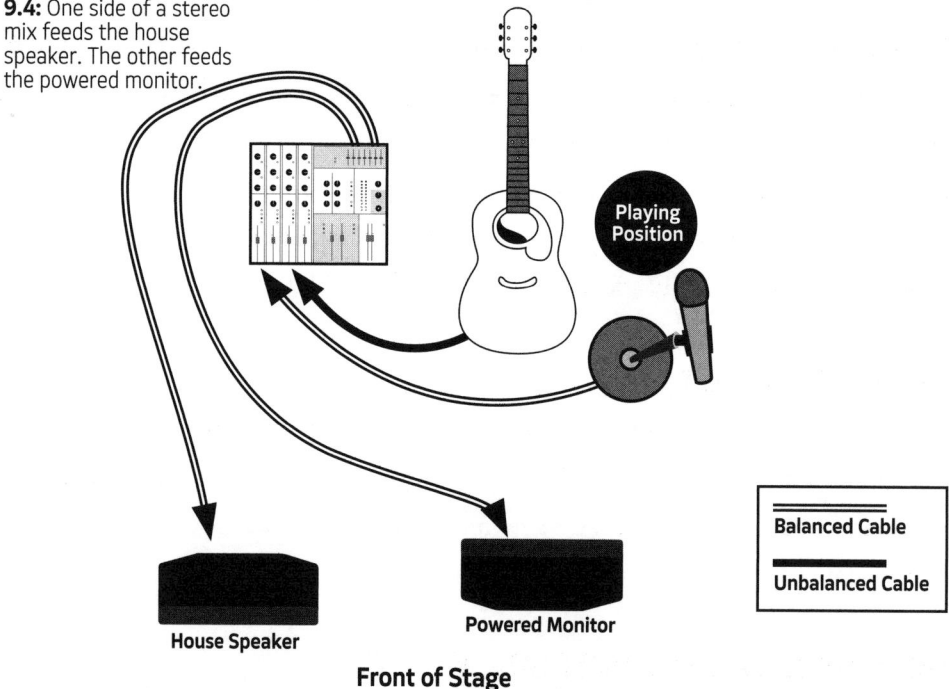

9.4: One side of a stereo mix feeds the house speaker. The other feeds the powered monitor.

Playing Position

Balanced Cable

Unbalanced Cable

House Speaker

Powered Monitor

Front of Stage

That said, you can create a similar setup by adding a single powered monitor or extension cabinet to an instrument amp, as shown in Figure 9.5. It all comes down to knowing how various components can work together!

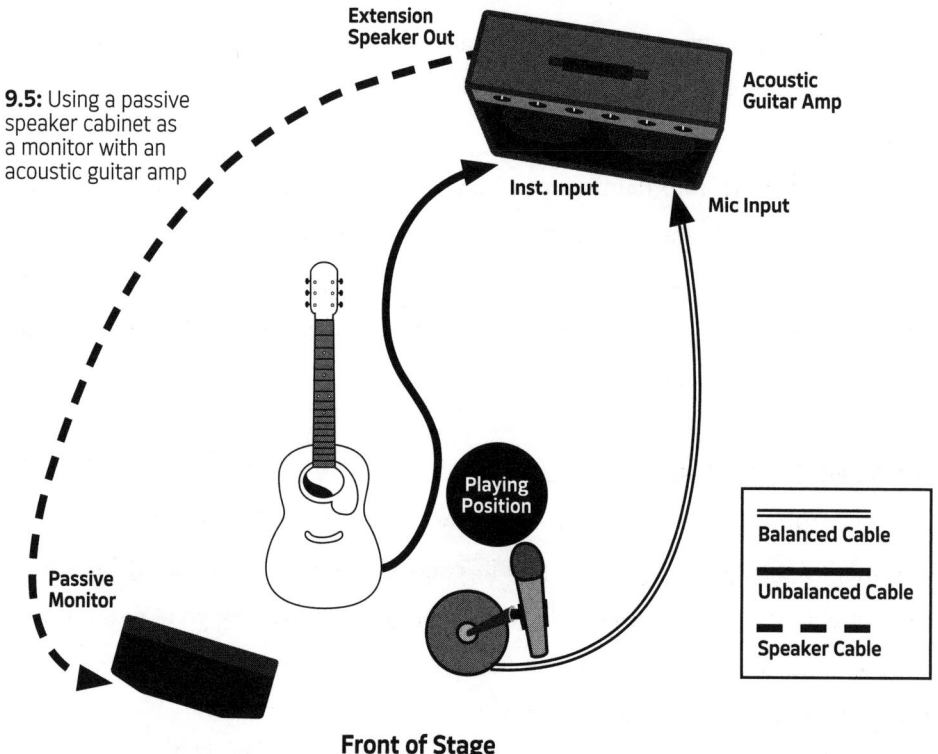

Extension Speaker Out

Acoustic Guitar Amp

9.5: Using a passive speaker cabinet as a monitor with an acoustic guitar amp

Inst. Input

Mic Input

Playing Position

Balanced Cable

Unbalanced Cable

Speaker Cable

Passive Monitor

Front of Stage

DUOS IN SMALL VENUES

The considerations for a duo or quiet trio are similar to those of a solo performer, with one exception: You need to be able to hear your partner as well as yourself. Members of a duo might each bring their own amps, share a P.A., or do a combination of the two. For example, in a duo with an acoustic guitarist and bassist, the guitarist might bring a small P.A. for vocals and guitar, and the bassist may bring a small amp for herself.

Figure 9.6 shows a simple setup for a duo. The players are in the middle; the bass player's amp is behind both performers; one P.A. speaker is in front facing the audience, and the other acts as a monitor for both players.

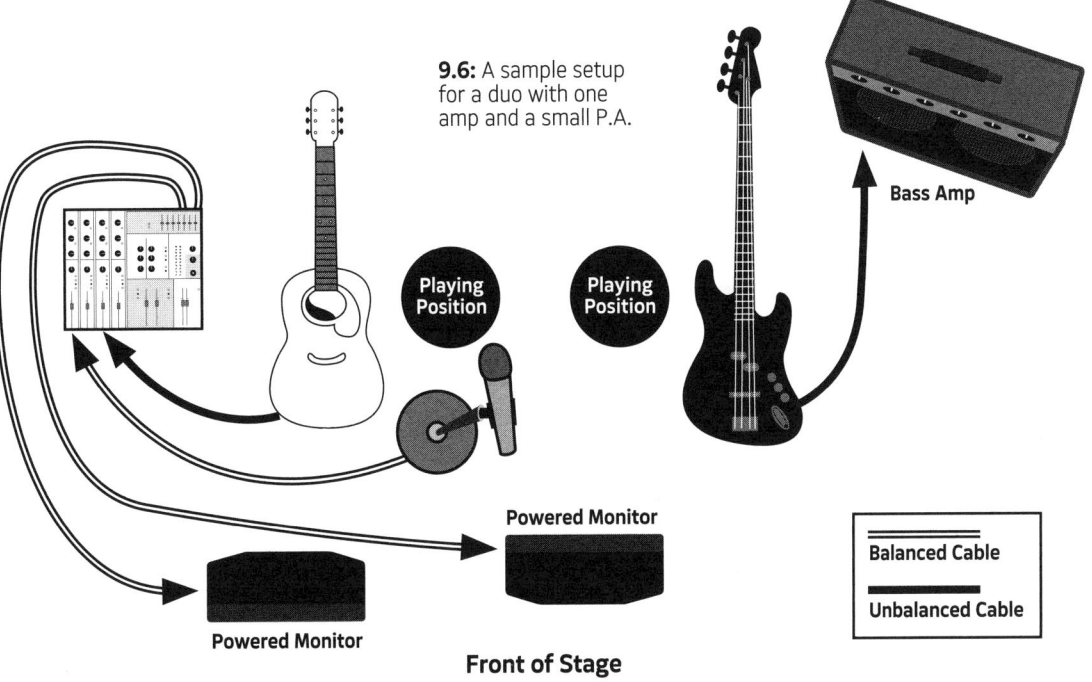

9.6: A sample setup for a duo with one amp and a small P.A.

Bass Amp

Playing Position

Playing Position

Powered Monitor

Powered Monitor

Front of Stage

Balanced Cable

Unbalanced Cable

Learning to Listen

It's really hard to get a good sound in a small space, and the main reason is because of volume. Amplified musicians play too loudly. We all do it! Part of the problem is that we can't hear ourselves among the other loud instruments. Another part of the problem is that we don't really hear what the audience hears.

You can try this exercise anywhere: Stand next to your amp and play. Now walk across the room and listen to it. You may discover that it's actually louder when you step away because of the way the sound projects from the speakers. You may also discover that the settings that seem perfect onstage sound muddy or overly bright when you listen from the audience.

A good monitor mix can solve this problem, but individual monitor mixes may be a luxury in a small

venue. It's never easy to play well when you don't like your sound, but you can train yourself during rehearsal to keep your own sound a little quieter and focus on the sound of your other bandmates. You have to trust that you sound good. You can also try moving your amp away from you on the stage, but raising it to ear level so that you can hear it better. When pointed toward you, your amp can act as a personal monitor.

Try to step out into the audience during soundcheck (Chapter 11) and listen from their vantage point as you and the whole band play. Finally, ask an objective listener to tell you if you're too loud or quiet. Don't ask your biggest personal fan. The club owner, a teacher, a member of another act, or anyone with decent ears can tell you if you need to turn up or down.

BANDS IN SMALL VENUES

A band playing a small venue faces a challenge: How do you get loud enough to compete with a drum set without being too loud for the room? In some cases, the drummer will have to play more quietly. Drummers can adjust their technique, use a sound barrier, or use "quiet" sticks, such as Pro-Mark Rods, which reduce the impact of a strike on the drumhead. (See Chapter 11 for more about drum sounds.)

In such a setting, the individuals will use their own amps—and therefore be responsible for their own mix—and the P.A. will be used mostly for vocals or acoustic instruments. Figure 9.7 shows a sample setup. Note that the P.A. now uses both main monitors for the audience, with separate stage monitors for the singers.

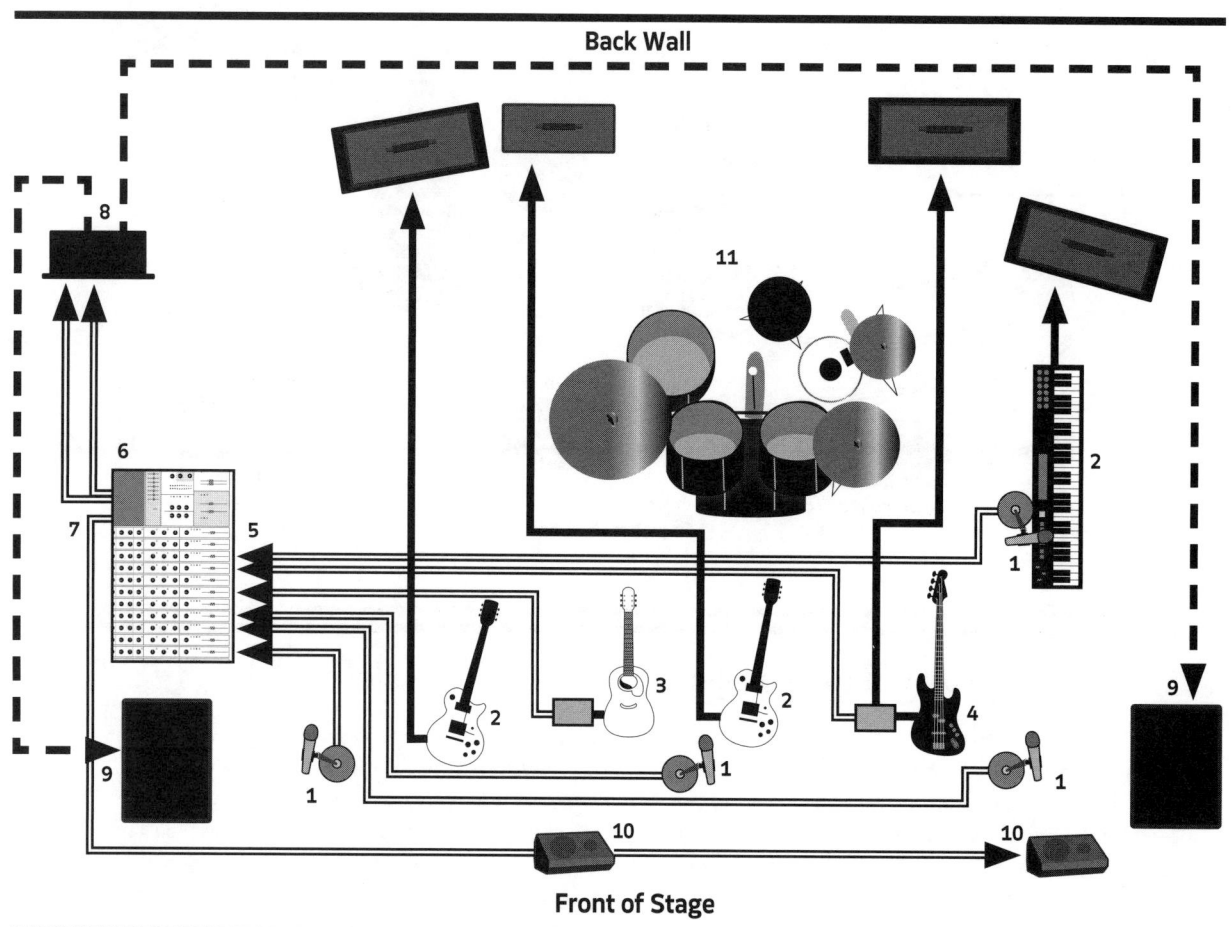

Balanced Cable

Unbalanced Cable

Speaker Cable

9.7: A small P.A. onstage with a combo band: 1) Microphones feeding mixer's mic inputs; 2) Instruments to amplifier inputs; 3) Instrument to direct box feeding mixer mic input; 4) Instrument connected to direct box feeding mixer mic input while passing signal through to amplifier; 5) Mixer microphone inputs; 6) Mixer main outputs (left and right) feeding power amplifier; 7) Mixer auxiliary output feeding powered monitors; 8) Power amplifier speaker outputs feeding house speakers (left and right); 9) House monitors sending main mix to audience; 10) Shared monitor mix for musicians onstage; 11) Unamplified drum set.

In this scenario, the band is mixing its own sound. Therefore the mixer—and all the cable runs—are onstage. You have to keep things neat and organized.

Midsized Venues

Larger venues require some additional considerations. First: How much (more) power will you need? If you have a small system, you may need to add more powerful amps and speakers—either through individual components, or by hooking up extra powered monitors.

Second: Will you need to use microphones on drums and amplifiers? This can make things much more complicated. Figure 9.8 shows two input lists for the same five-piece rock band in two different settings: one requiring the P.A. for vocals and acoustic instruments, the other using the P.A. for amps and drums as well. We'll discuss specific miking techniques in Chapter 10, but for now, let's just assume we're using one mic per instrument.

SOURCE	Small Venue: Minimal P.A.	Midsized Venue: Full P.A.
Singers (4)	4 Mics	4 Mics
Guitar Amp 1	0	1 Mic
Guitar Amp 2	0	1 Mic
Bass Amp	0	1 (D.I.)
Acoustic Guitar	1 (D.I.)	1 (D.I.)
Keyboard Amp	0	2 (D.I.)
Drum set (kick, snare, three toms, overheads)	0	7 Mics
Total Mics	4	13
Total Inputs (including D.I.s)	5	17

9.8: Counting up the inputs you'll need can help you prepare for shows at venues of various sizes.

The list above doesn't even account for things like monitor mixes, effects, etc., but it does tell you that you'll need a mixer with at least 17 inputs. Since mixers tend to offer inputs in multiples of four or eight, your choice is between using a 16-input mixer and dropping one or two of the mics or D.I.s (for example, running the keyboard only through the amp or a mono D.I., or forgoing one or both of the drum overheads) or using a 24-input mixer. If you know your mixer's buses, you could also get creative and use one of the mixer's auxiliary returns to handle the keyboard, since its output level is a good match for a typical aux input. Figure 9.9 shows this option. For the first time, we've moved the mixer off the stage to a mixing position. To keep the cable runs neater, we're using a snake.

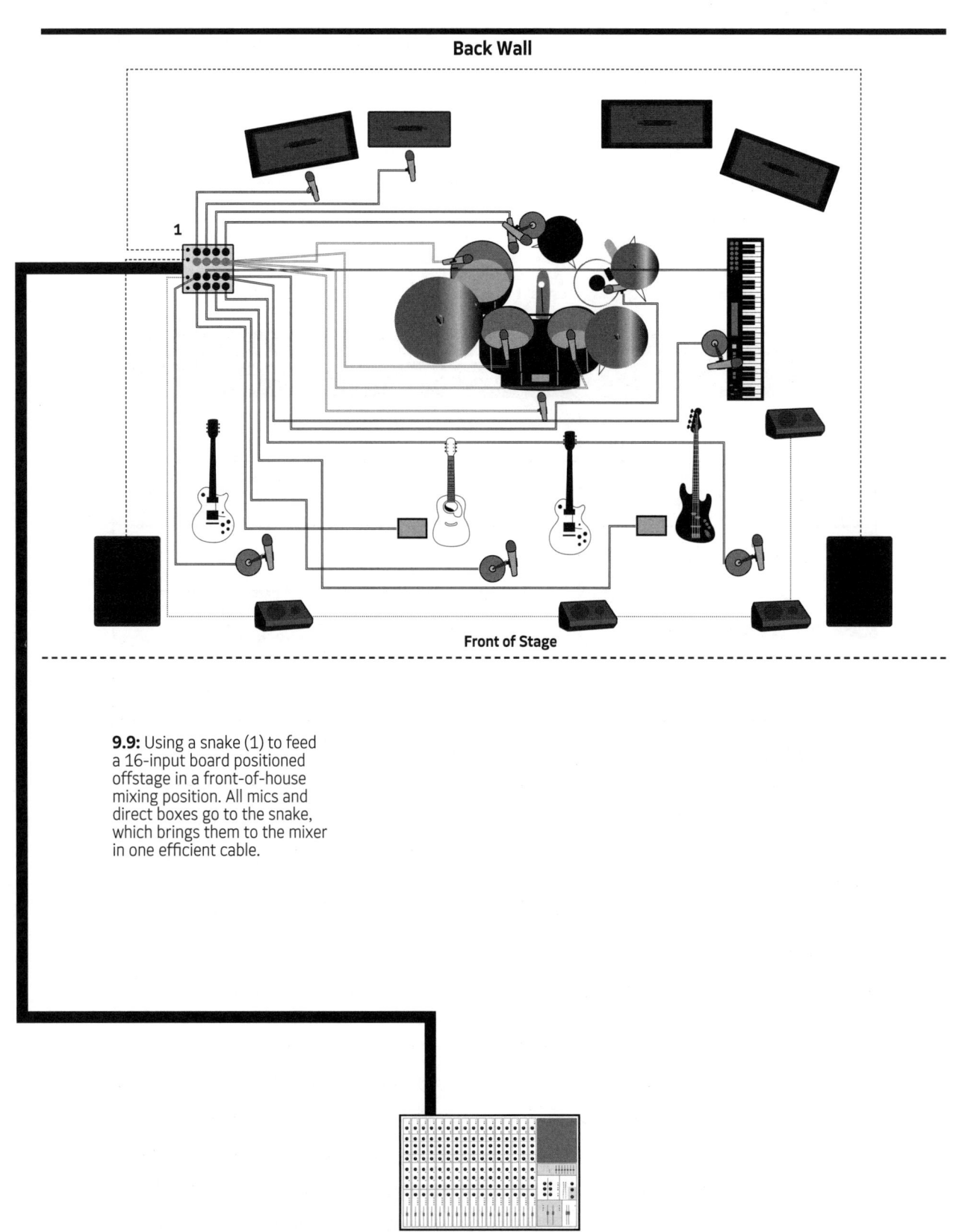

9.9: Using a snake (1) to feed a 16-input board positioned offstage in a front-of-house mixing position. All mics and direct boxes go to the snake, which brings them to the mixer in one efficient cable.

Setting Up an Offstage Mix Position

So far, we've been focused on the stage, but as you can see in Figure 9.9, there are times when a mixing position is away from the stage. For this, you'll need to have one (or more) people who will work on sound during the entire show. The ideal position is some distance from the stage—it can be in the back of the room or in the middle (which will give you a more accurate picture of the sound). If possible, set up so that your listening position is facing the center of the stage, so that you can get a perspective on the entire mix. If possible, position the mixer twice as far from the stage as the left and right house speakers are from one another (i.e., if the speakers are 20 feet apart, set up 40 feet from the stage).

RUNNING SIGNAL TO AND FROM FRONT OF HOUSE

When the mixer is that far from the stage, you'll be making longer cable runs. In Figure 9.9, we showed an audio snake. These are almost essential if the mixer is more than a few feet from the stage. (You could get by without one, but the setup would be unwieldy.)

Let's get into a little more detail. The stage side of the snake has XLR inputs for microphones and direct boxes. On the mixer side, the snake breaks out into separate wires, each with its own XLR connector that can plug into the mixer's mic inputs.

But a snake can run signal *to* the stage as well. Four 1/4" jacks on the snake can be used to send signal from the mixer to the amps driving stage monitors, front-of-house speakers, or other destinations. Figure 9.10 shows how things might be connected to a 16 x 4 snake.

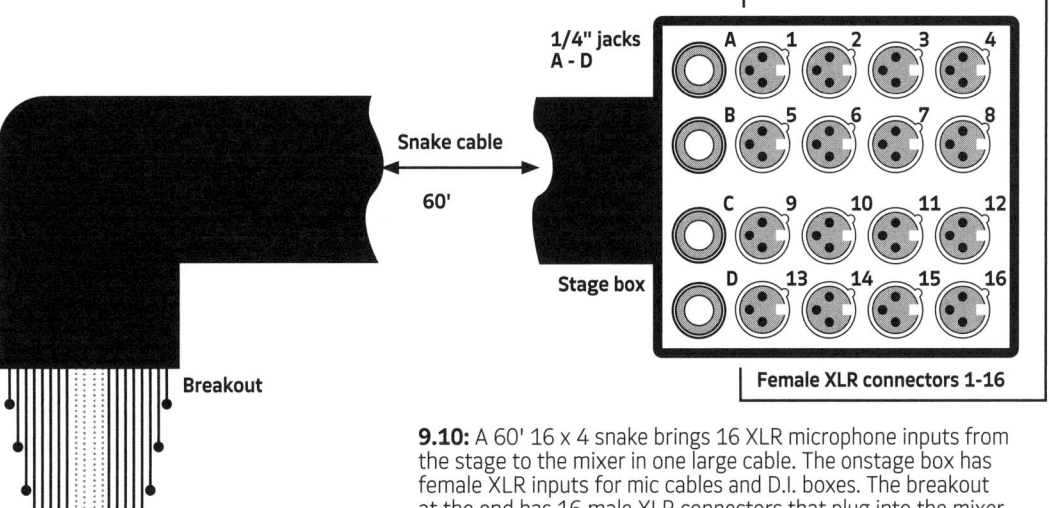

9.10: A 60' 16 x 4 snake brings 16 XLR microphone inputs from the stage to the mixer in one large cable. The onstage box has female XLR inputs for mic cables and D.I. boxes. The breakout at the end has 16 male XLR connectors that plug into the mixer (or other mic preamp). The snake's four 1/4" connectors can be used to send signal from the mixer to the stage or to send line-level signals from the stage to the mixer. For example, here's where the connections shown in Figure 9.9 went: 1) Guitar amp left; 2) Guitar amp center; 3) Drum overhead left; 4) Drum overhead right; 5) Floor tom; 6) Kick drum; 7) Rack tom left; 8) Rack tom right; 9) Left guitarist's vocal mic; 10) Keyboard line out; 11) Keyboardist's vocal mic; 12) Snare drum; 13) Acoustic guitar D.I.; 14) Center guitarist's vocal mic; 15) Bass D.I.; 16) Bassist's vocal mic; A) Main mix right to powered house speaker right; B) Main mix left to powered house speaker left; C) Auxiliary send 1 to stage monitors; D) Auxiliary send 2 (not connected).

Be Prepared!
Pro Tip by Kevin Madigan, Front-of-House Engineer

Before you do anything, read and understand as much as you can about sound systems and their interaction with their environment. Here are some good resources: *Sound Systems: Design and Optimization: Modern Techniques and Tools for Sound System Design and Alignment* by Bob McCarthy; *Sound System Engineering* by Don and Carolyn Davis; and the *Sound System Design Reference Manual*. The latter is available for free from the pro audio equipment company JBL at jblpro.com. (See disc for a direct link.)

1. PREPARE
- Do you have a channel list? A simple stage plot? If not, write some up.
- Find out what your available resources are: P.A. monitors, microphones, console.
- Are you familiar with the console? If not, download the manual. You may even be able to prepare your show with "offline" software for a digital console.

2. SET UP
- Do you have everything on your channel list?
- Listen to your P.A. Learn about P.A. optimization and implement your knowledge.
- Decide—and be realistic—about what level you want to run your show at.
- Zero the console if it's not already done.

- Choose your instrument and microphone positions carefully.

3. SOUNDCHECK
- Check that each line is working and get an approximate gain level with a safe margin before clipping and before feedback.
- Listen. Is *that* how you want that to sound? Would it be better to move the mic than change the EQ?

On many consoles you'll find a hi-pass filter. The frequency of this control may be variable, or it may be preset to a single frequency (such as 100Hz). Using it on channels such as vocals will help eliminate low-frequency feedback. (See Chapter 11 for more on soundchecks and zeroing the mixing board.)

MIXERS TO POWER AMPS TO SPEAKERS

Between front-of-house and monitors, you're going to be connecting the mixer to at least two (and maybe quite a few more) channels of power amp. Remember that a stereo power amplifier can send a mono signal to each of two separate destinations, so you can use one side for front-of-house, the other for monitors.

Power amps can go one of two places: near the mixing board or near the monitors. Of course, if you have a powered mixer, the power amp is already so near the mixer that it's housed in the same structure. Likewise, if you have a powered speaker, the amp and speaker share housing.

If you're using a separate power amp, you have to decide which cable you want longer: the speaker cable running between the amp and the cabinet, or the audio cable running between the mixer and the amp.

If you're running a balanced signal from the mixer to the amp, you're probably better off positioning the amps near the speakers and using as short a speaker wire as possible.

COMMUNICATING WITH THE SOUND MIXER

When the people doing the mixing are far away from the musicians, they're not going to be able to talk very easily. Of course, the musicians can use the mics onstage to talk to the sound mixer. But in order for the mixer to talk to the musicians, he or she can use a microphone called a *talkback* mic that's plugged into the board but routed only to the onstage monitors, not to the house system. This is easy to do if you're using aux inputs for the monitor mix. Just be sure to turn the mic off when you're not talking to the performers, so that their monitor mix isn't cluttered by sound from front-of-house. Some mixing boards have talkback mics built in.

Hand signals are also an effective form of communication, especially if you work them out ahead of time. For example, let's say you want to hear your vocal louder in your monitor, but don't want to say so into the mic in the middle of a performance. Point to yourself, then your monitor, and then point up. If the sound mixer is watching, he'll get the message.

Working With an Audio Professional

In all of the scenarios above, we've assumed that you or someone with you will be setting up and operating the sound. But any venue, even a small one, may already have a sound system in place and a person to operate it. In that case, you need to adapt your own setup to fit the existing system.

In a larger venue, it's very likely that you'll be working with the location's sound setup. If you have your own sound engineer, he or she will need to get familiar with the existing equipment. If you're working with the house engineer, the steps we outlined in the previous chapter—sending a stage plot with input list, following up with a phone call to go over what you need—will help.

YOUR BACKLINE WITH THEIR P.A.

We've talked about situations where the venue provides a full backline, but that's not always the case. Here's one scenario that young musicians might experience: playing a town or county fair with an outdoor stage. The organizers may have some equipment in place, like a drum kit, but you'll probably have to bring your own amps and set them up, decide what to mic and what to run without mics, etc.

If you're using your own amps, do yourself a favor and let the sound engineer set the volume. Get your basic sound and step back. If you can't hear yourself, ask for more guitar or bass in the monitor, but don't turn the amp up. By leaving it quiet onstage, you actually give the sound pro more of a chance to boost your amp with the P.A., control its mix in relationship to the other instruments, and let the audience hear you more clearly.

Monitoring

Throughout this book, we've made reference to onstage monitoring. Monitors allow the musicians to hear themselves and each other. In large touring setups, every player can have his or her own monitor mix. In fact, there may be a sound mixer whose sole job is the monitor mix. In other cases, monitor mixes can be a compromise. You may get one monitor mix for everyone, or two monitor mixes: one for the lead singer, the other for everyone else.

POSITIONING MONITORS

Typically, stage monitors should go on the floor, pointing up and facing the performer. Figure 9.11 shows monitors at each microphone position at the front of the stage. The monitors are behind the mics, pointing at the player, thus reducing the possibility of feedback.

Sometimes, two players will have to share a single monitor. In this case, it can be positioned between the two mics, as shown in Figure 9.12.

9.11: Each player uses his own monitor. The shaded area shows where the speakers project sound.

9.12: Positioning a monitor so that it can be shared by two musicians

The monitor moves left and the players move closer together to share its sound.

Back Wall

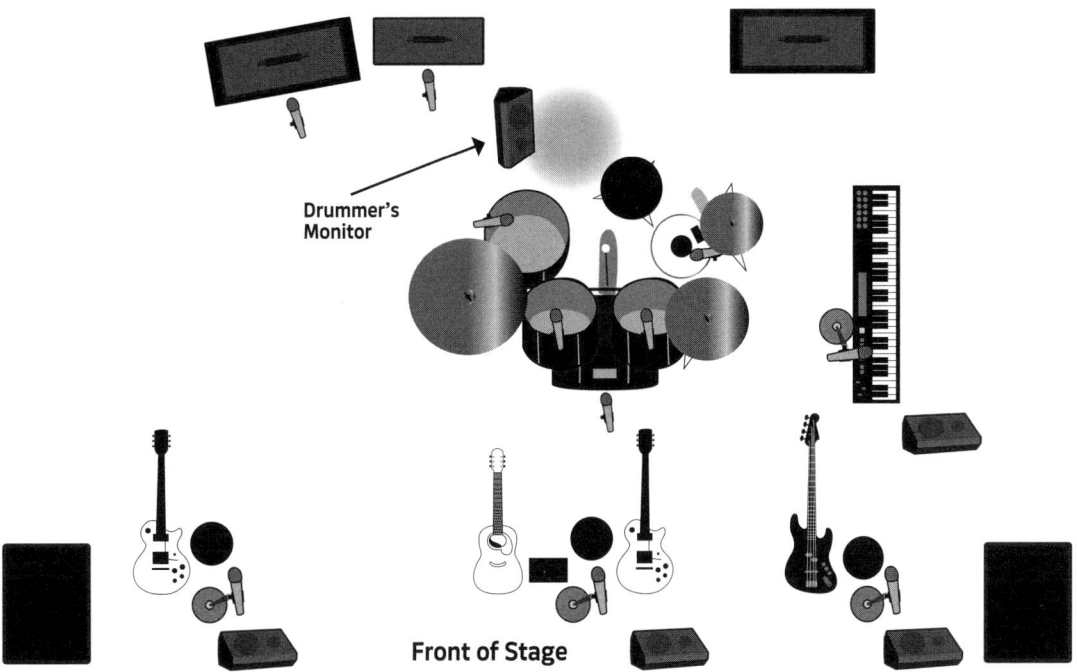

Drummer's Monitor

Front of Stage

An upward-facing wedge isn't as effective for drummers. Instead, you can use an upright cabinet, pointed at the drummer from the side, as shown in Figure 9.13. Just be sure to keep any drum mics out of this monitor mix.

Using Stage Amps as Monitors

As I mentioned earlier, you can use your own amp as a monitor by positioning it where you can hear. This can actually be pretty useful in large venues where the amp is being miked or an instrument is going direct. Since the amp isn't feeding the house, you can move it as needed. Acoustic guitar amps, for example, make great stage monitors if the house is taking a direct input from the guitar. Turn the amp to face you, and run a signal from the guitar to

Cable Runs
Pro Tip by Eric Turquman, Front-of-House Engineer

Usually one would prefer to send a balanced signal from the mixer to the amp, with the amp at the stage—provided your mixer sends out a balanced signal and the amp can accept it. However, as long as you aren't pushing a heavy-duty signal (such as powering subwoofers), 40 feet isn't really going to make a difference and the amp could be placed by the mixing desk. The best reason for having a power amp by the stage is that it's usually the shortest run to the electrical service. You always want to minimize the distance that the electrical power has to run to the amp. Longer AC runs to amplifiers often can lead to tripping circuit-breakers—especially if you're using lighter-gauge electrical wiring.

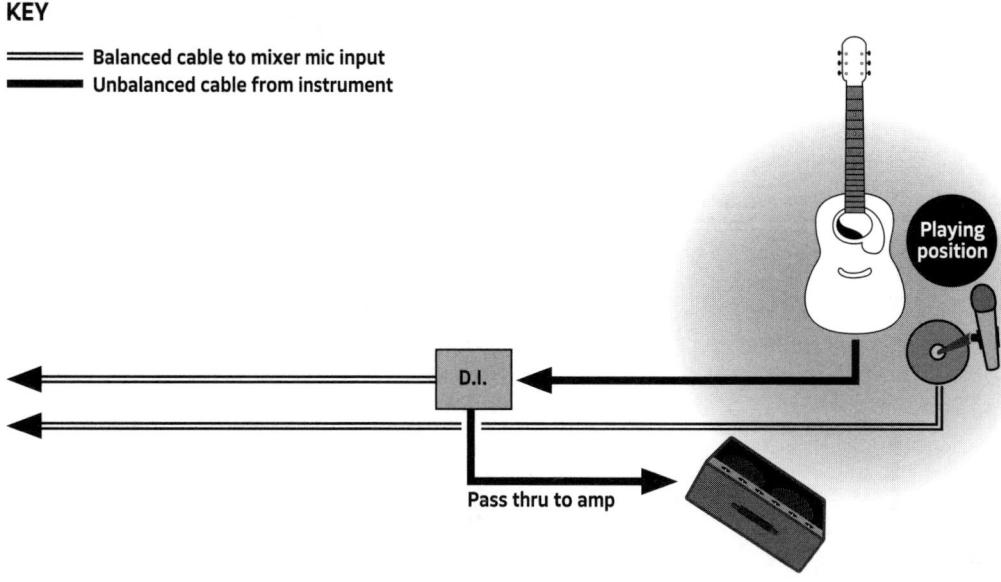

9.14: Using a direct box to feed the house, and using an onstage amp as a personal monitor

KEY

══════ Balanced cable to mixer mic input
▬▬▬▬▬ Unbalanced cable from instrument

Playing position

D.I.

Pass thru to amp

the direct box, an unbalanced "through" line from the direct box to the amp, and a balanced line from the direct box to the house system (Figure 9.14). This scenario also works well for keyboards and bass. With electric guitar, the amp is more likely to be miked and positioned in the backline.

ROUTING SIGNAL TO MONITORS

With a small system or a very simple mix, the monitors may get the same sounds as the house system. But with a more complex setup, the monitor mix and the house mix might be very different. We'll discuss some monitor mix strategies in Chapter 11, but for now, let's keep in mind that every separate monitor mix will require its own connection, amplification source, and speaker (or in-ear receiver). Figure 9.15 shows an elaborate setup, where each musician receives his or her own monitor mix.

Wireless In-Ear Monitors

Wireless in-ear monitors allow players to hear themselves no matter where they are onstage. They can also control the level of their monitor by adjusting it at the receiver, though they won't be able to control the balance of the monitor mix itself.

Instead of running a cable to the stage, the cable runs from the mixer to the wireless transmitter, which in turn feeds a receiver worn by the players. Figure 9.16 shows the same arrangement as Figure 9.15, replacing the onstage monitors with a wireless system.

9.15: Using cables to send monitor mixes from the mixer to the stage

Back Wall

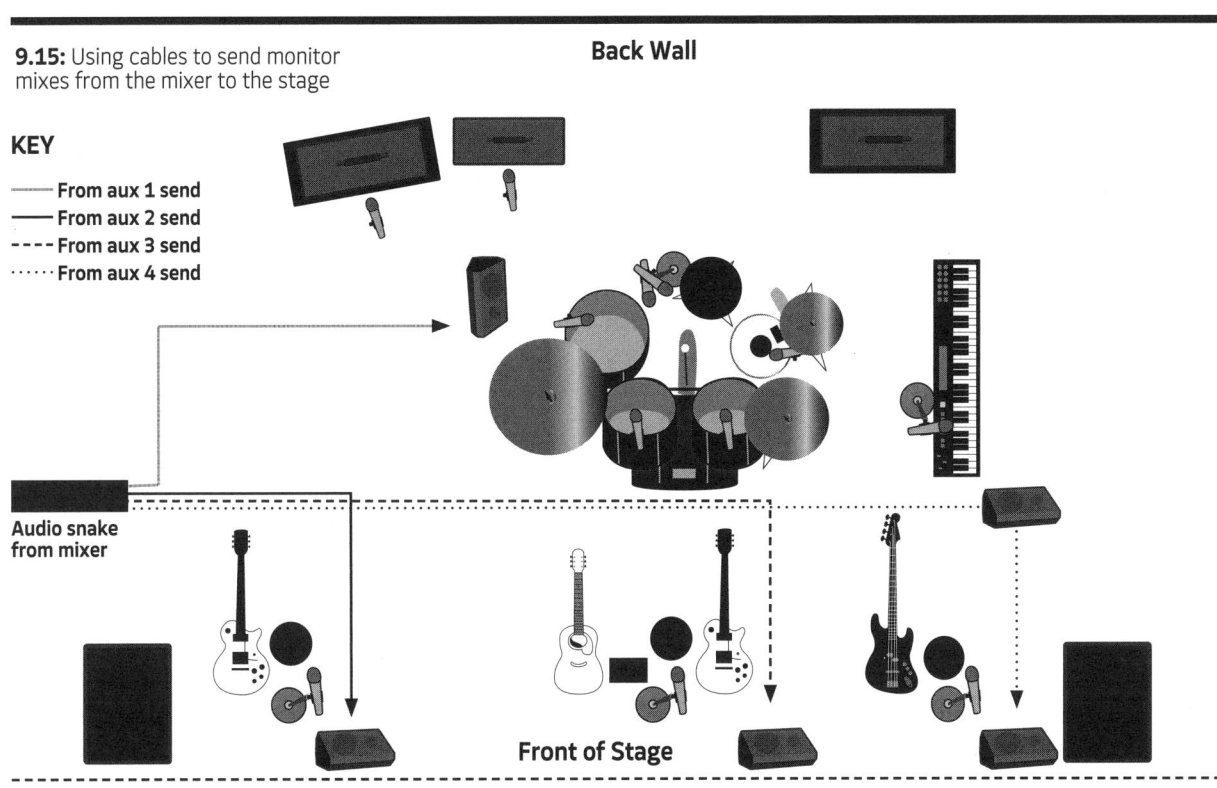

KEY

──── From aux 1 send
──── From aux 2 send
- - - - From aux 3 send
· · · · · From aux 4 send

Audio snake from mixer

Front of Stage

9.16: Using a wireless system to send monitor mixes from the mixer to the stage

Back Wall

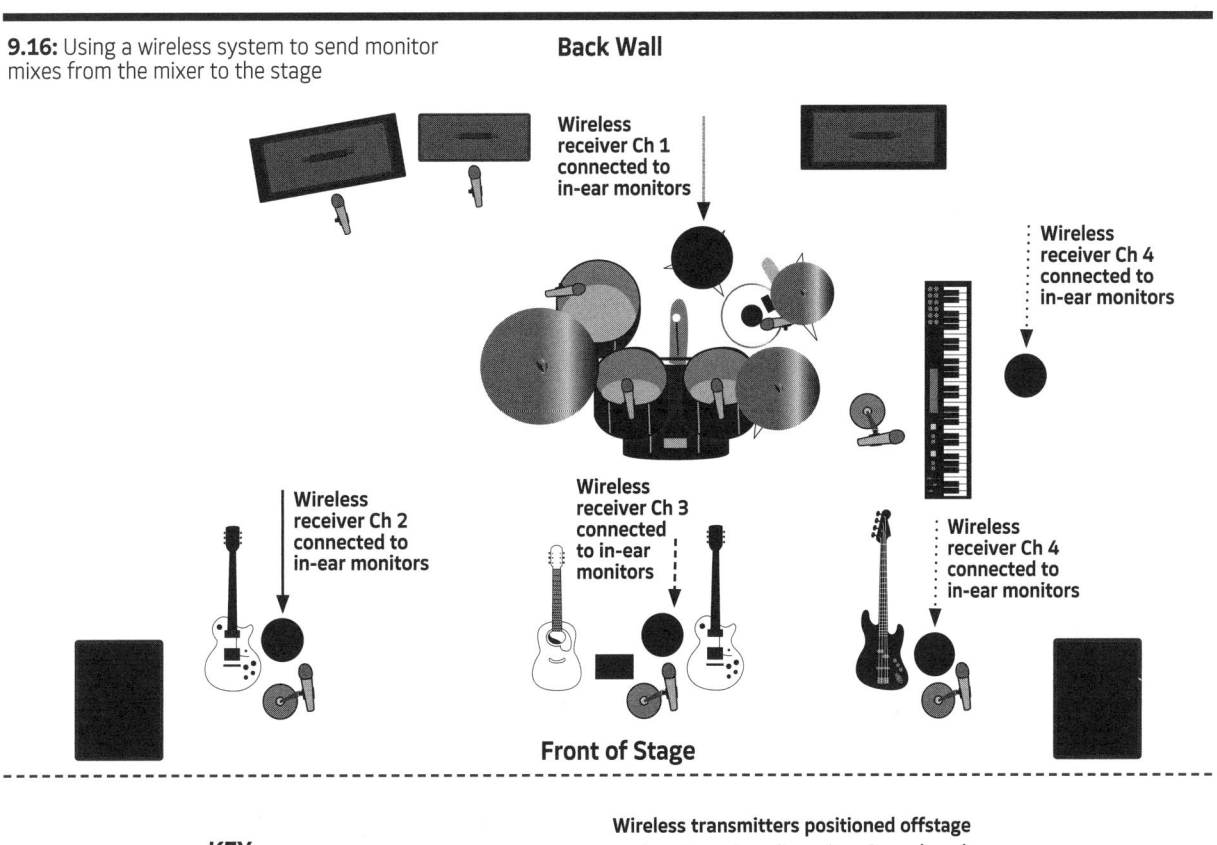

Wireless receiver Ch 1 connected to in-ear monitors

Wireless receiver Ch 4 connected to in-ear monitors

Wireless receiver Ch 2 connected to in-ear monitors

Wireless receiver Ch 3 connected to in-ear monitors

Wireless receiver Ch 4 connected to in-ear monitors

Front of Stage

KEY

──── From aux 1 send
──── From aux 2 send
- - - - From aux 3 send
· · · · · From aux 4 send

Wireless transmitters positioned offstage

| Ch 1 | Ch 2 | Ch 3 | Ch 4 |

Audio cables from mixer

Signal Processors

Effects and signal processors should be positioned where they're easiest to control. Those used by the performers onstage should be within their reach, either on pedalboards or in racks near the individual's amplifier. For complex rigs, a switching system or controller can make things easier.

Any signal processors used as part of the main mix should be positioned close to the mixing console. If you have a mixer mounted in a rack (see Chapter 7), you may have already pre-wired effects to the mixer itself, or to a patch bay connected to the mixer and stage snake. The ideal setup allows you to operate the signal processor's controls without leaving your mixing position. Figure 9.17 shows a setup with the mixer facing the stage and the signal processors in a separate rack positioned to the mixer's side in an L shape.

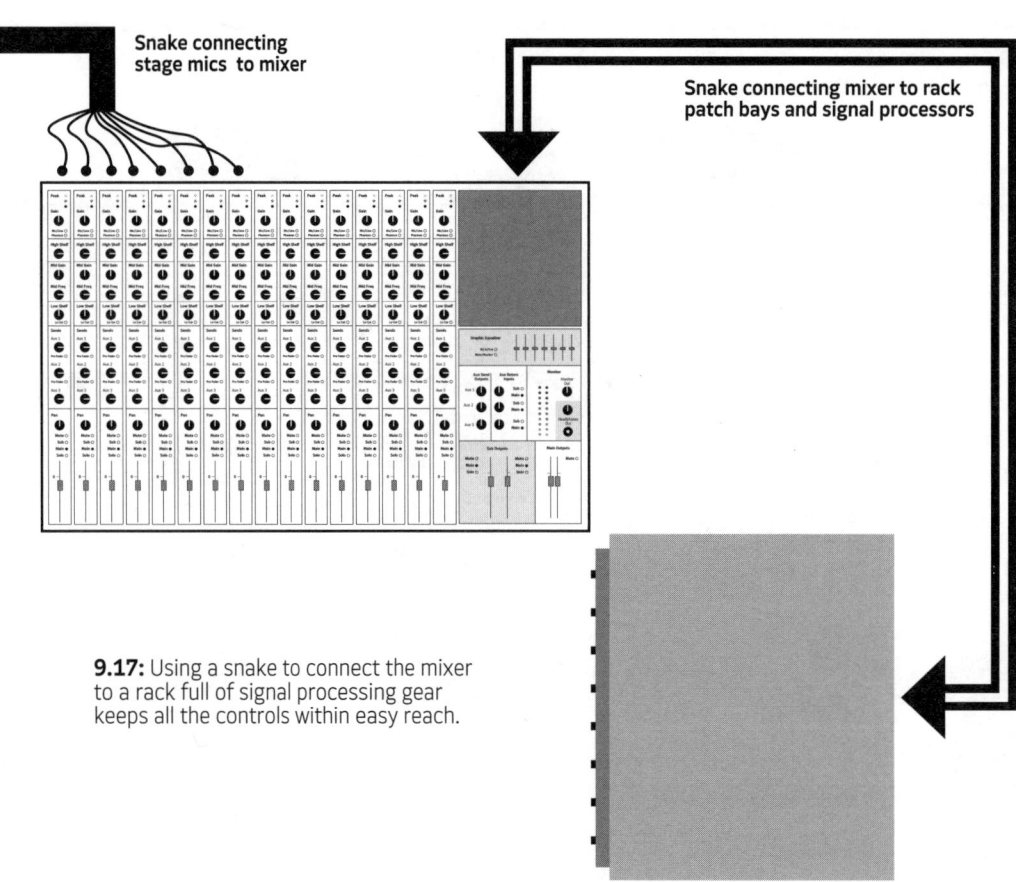

Snake connecting
stage mics to mixer

Snake connecting mixer to rack
patch bays and signal processors

9.17: Using a snake to connect the mixer to a rack full of signal processing gear keeps all the controls within easy reach.

Onstage Submixes

We discussed the concept of submixes in Chapter 2. A submix is a group of sounds that are mixed so that they're balanced with one another, then routed so that they can be controlled as a unit within a larger mix.

When a stage performer has a complex setup with lots of inputs, it may be helpful to mix them together using a small onstage mixer, then send that submix to the house system. The mixer used for such a task can be simple. A device called a line mixer—which has no microphone inputs—is a common choice. It can fit in the same rack as a player's onstage effects and sound modules. Figure 9.18 shows how a small onstage mixer might be used to consolidate multiple inputs from a keyboardist's rig, then send a stereo feed to the house mixer.

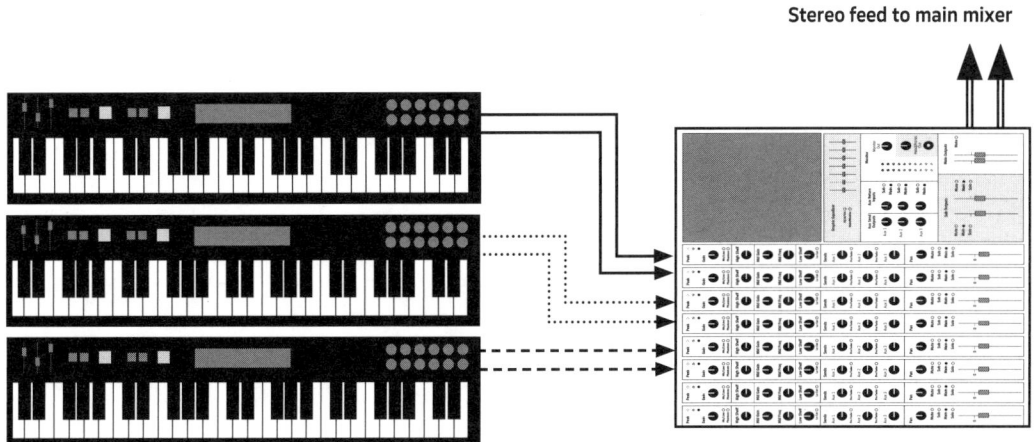

9.18: These three keyboards have a total of six outputs. A small onstage mixer can consolidate them into a stereo signal that feeds the house mixer. This also allows the keyboardist to tweak the balance between the instruments during the show.

Next Stages

There are many ways to set up a P.A. system, and it's important to be able to adapt to the circumstances at hand. Performers may never actually touch the P.A.'s controls, but if they understand what the sound mixer's trying to accomplish, they'll work together more effectively. In the next chapter, we'll look at how to set up and use microphones and pickups to bring vocal and instrument sounds from the stage to your mix.

Chapter 10

SETUPS FOR MICROPHONES AND PICKUPS

Now that the mixer is set up, it's time to get some sound into it. In this chapter, we'll look at options for capturing voices as well as acoustic and electric instruments using microphones and pickups.

In Chapter 4, we discussed types of microphones and pickups and their applications in live sound. Now we'll put them to use on the human voice, various instruments, amplifiers, and whole ensembles. Choosing the right microphone is only half the battle. Positioning the mic is also important.

In some cases, a pickup is a better choice than a mic. As we explained in Chapter 4, pickups are devices that convert vibrations into electrical energy. There are many varieties, but two basic types: *magnetic pickups*, which read string vibration, and *contact pickups*, which read the vibration of a surface (such as the bridge of a guitar). Electric guitars and basses typically use magnetic pickups, while contact pickups are often used with acoustic instruments. Some pickups actually have microphone elements as well.

Before reading further, you may want to refer to the sections of Chapter 4 that deal with a microphone's pickup pattern, frequency response, sensitivity, and proximity. No matter what kind of mic you use, you should read its documentation carefully so that you know how that particular mic behaves. Never just assume you know how an unfamiliar mic will sound. Just because two different models of microphone may look similar, that doesn't mean they behave the same way.

Plugging in a Microphone or Pickup

Sound gets from the stage to the audience in one of three ways. There's the ambient sound produced by instruments; there's the sound captured by microphones; and there's the sound captured by pickups.

In some cases, the mics and pickups will be simply reinforcing the ambient sound of the ensemble. An orchestra or big band needs very little help filling a small auditorium. Sometimes, the mics and pickups will help individual instruments stand out among louder ambient instruments. Sometimes, they'll offer the only way for a sound to reach the audience (an electric guitar, for example, would be inaudible without its pickups).

In all these cases, the microphone or pickup will only be effective if it's sent to the right destination in your signal chain.

MICROPHONE INPUTS

Microphones put out a relatively weak audio signal. This signal needs to be boosted by a preamp before it's ready to be amplified and sent to the house and stage systems. As we discussed in Chapter 2, mixers have mic preamps built into them. You'll also find standalone mic preamps, as well as some instrument amplifiers that have microphone inputs.

Try to use these inputs at all times. Avoid plugging a mic into the instrument inputs on an amplifier (there are some exceptions to this rule, such as running a bullet mic into an amp for harmonica). Mics can plug directly into a mixer/preamp or into an onstage snake that runs to the mixer/preamp.

PICKUPS

Most pickups put out a weak signal on their own, so they need to go to instrument-level or mic-level inputs. When the mic inputs are used, the pickups' unbalanced output should be converted to a balanced signal. You do this by plugging the pickup into a direct box, which gives the signal the proper level and impedance to feed the mixer's microphone input.

There are several different ways to use a pickup. You can connect it to an amplifier on stage and bypass the P.A. entirely. You can route the pickup to an onstage amp that will be miked and fed through the P.A. You can connect it to a direct box (D.I. for short) and feed the P.A. system from the stage without using any other amplification. You can run the pickup to both a D.I. and to an onstage amplifier. Or you can send the pickup to a specially designed preamp, and route the preamp's signal to the mixer's line inputs. Which is best? That depends on the instrument and the sound you're going for.

ELECTRONIC INSTRUMENTS

Electronic instruments, such as synthesizers, samplers, electronic drum sets, and drum machines, don't use pickups. Their circuits are fed to internal preamps, which route sound to their outputs. Many electronic instruments actually have built-in mixers.

Like pickups, the signal from electronic instruments can be fed to an amplifier or directly to a mixer. The big difference is that some electronic instruments can produce a signal that's strong enough to go into a mixer's line-level inputs. Some electronic instruments also have digital outputs that can connect to a digital mixer's digital inputs.

Computers

For the purposes of live sound, you can think of a computer as an electronic instrument. The kind of signal it produces will depend on the audio interface you use. The computer's built-in soundcard will most likely offer an unbalanced stereo feed that's really designed to drive headphones. An external interface, which connects to the computer by USB or Firewire or is mounted as an internal sound card, may have unbalanced or balanced outputs. It will most likely offer both analog and digital outs.

TAKING A DIRECT FEED FROM AN AMPLIFIER OR PREAMP

If your amp has a direct output, you can send that to the mixing board. The line output sound may seem brighter than the amp itself as it feeds its own speaker. This is because the speaker's influence is removed from the signal path. Some amps offer speaker simulation at the direct out, which makes this feed sound more like the amp running through a cabinet.

Guitar and bass preamps can also be used in this way. Most models have built-in speaker simulation. Digital preamps may even let you choose the speaker cabinet and microphones being used in the simulation!

If you take a direct feed from an amp, use a D.I. box—even if the output is at line level. Sometimes, you may discover a loud hum when the amp is plugged into the board; this may be caused by a ground loop, and the D.I. ground lift switch—which lifts the audio ground, not the electrical ground—can solve the problem.

Vocals

Probably the most common task for the P.A. will be amplifying vocals. Even in a small venue where the instruments are unmiked, the singer may need a microphone to be heard.

Singing into a microphone seems to be the simplest thing in the world. Step up to the little round part and start vocalizing. So simple, even a singer can do it! Okay, bad jokes aside, if singing into a mic is so simple, why do live vocals sound so bad so often?

There are four main culprits, and they usually work hand in hand:

1. You're not in the right position for the part you're singing—for example, pulling away from the mic on quiet passages, coming in too close on loud passages, or moving your mouth away from the mic's sweet spot for any other reason.

2. The mixer and/or amplifier settings are making the mic sound worse than it needs to.

3. You're screaming (or singing too quietly). Usually the former. Often, this is a result of problem 2.

4. You're singing into a mic that's poorly suited to vocals—for example, it may produce a lot of handling noise or be prone to pops and feedback. We'll tackle the last problem first.

CHOOSING A VOCAL MIC

What makes a microphone a "vocal mic?" That's a good question. A singer can sound fine on the same microphones that people use for guitar amps, kick drums, and the like. That said, most manufacturers describe specific models as being particularly suited to vocals. The Shure SM58 (Figure 10.1) is a prime example. It's actually a very good mic on instruments, closely related to the Shure SM57, but its capsule is slightly more isolated and has a built-in windscreen.

10.1: Shure SM58 with the cap unscrewed, showing the element (left) and windscreen (right).

A vocal mic usually has a built-in pop filter to help reduce those "popped P" sounds that are known as 'plosions. The mic may have some internal shock absorbers to help isolate the mic's element—the part that picks up sound—from the body, in order to reduce handling noise. Finally, a vocal mic for stage use should also have a cardioid or hypercardioid pickup pattern to help isolate it from other onstage sounds and reduce feedback. We discussed these factors in some detail in Chapter 4.

As we mentioned in that chapter, dynamic mics are more common for stage vocals than condenser and ribbon mics. They're generally hardier, which is important because mics get a lot of abuse on stage. Also, because they're generally less sensitive, dynamic mics are less prone to feeding back and picking up unwanted sounds.

That said, there are a few condenser mics that are built specifically for stage use, such as the Neumann KMS140 shown in Figure 10.2. These are very different from the kind of condensers you'd find in a recording studio, so you can use them as you would use a dynamic mic. However, stage condensers can be a lot more expensive than their dynamic counterparts. Unless you have deep pockets, you may want to start with a dynamic.

10.2: Neumann KMS140

Finding What Suits Your Voice

There are dozens of good vocal mics on the market, ranging in price from less than $100 to several times that much. What should you choose? The answer has as much to do with the way you sing as it does with the mic. Some singers feel that they sound much better with a particular model—perhaps it brings out a little more of the highs, or emphasizes the midrange in a different way. The only way to tell is to try a bunch of mics for yourself and listen. If you can record the sound, so much the better.

Get a Grip

Sound may be the most important factor, but if you're going to be holding the mic for hours on stage, it should feel good in your hands. So when you shop for a mic, pick up a bunch of different models and imagine you're singing through them.

Switch or No Switch

There was a time not long ago when any mic that had an on-off switch seemed like a toy, or something designed for amateurs at best. After all, serious performers plug into mixers, and—as we discussed in Chapter 2—mixers have mute switches! But due to the growing popularity of portable P.A. systems and microphone-friendly amps (such as those built for acoustic guitars and keyboards), it makes sense to give a solo performer the ability to mute and unmute his or her own mic onstage.

Manufacturers have started to add switches to their pro models; you can see an example on a Shure SM58 in Figure 10.3.

10.3: Vocal mic with an off switch

MICROPHONE POSITION

Good musicians know that posture, hand position, embouchure, etc., are all essential to producing good playing technique. The same rule holds true when you're interacting with technology.

For singers, mastering the art of microphone position is almost as important as learning to breathe and support your voice. The next time you watch a really great singer on stage (or on TV), pay attention to how he or she holds the microphone. You'll probably notice that on some notes, the mic is farther from the singer's mouth, and it may be pointed at a little bit of an angle. For other notes, the singer's lips are only inches away from the mic—or even closer. If you're a really keen observer, you may even note that the mic never seems to get between the singer's face and the camera.

Singing into a mic with such precise attention to detail is far from natural; it takes practice and discipline. And as you do with scales and breathing exercises, you should work on your microphone technique at home and in rehearsal, so that you don't have to think about it much on stage.

After all, a singer is supposed to be delivering an emotional performance. If you're singing about heartbreak, thoughts like "the ideal position for the next phrase is 11 centimeters from my mouth" don't generally come to mind. With practice, the mic becomes part of your body, and the way you position it becomes part of the gestures you use to help reinforce the song's meaning.

Handheld Mics

Singers using handheld microphones have the ability to control the mic position simply by moving the microphone. Figure 10.4 shows singer Heather Rose from the Lagond Music School in Elmsford, New York, holding a mic in two different positions: close for quiet passages, and farther away for louder passages.

10.4: Close and more distant mic positions

As you learn what works for you, you can transfer some of that knowledge to using a mic while it's mounted on a stand. Therefore, I recommend that every singer practice with a handheld mic—even instrumentalists who always sing into a stand-mounted microphone when they perform.

Always be aware of your surroundings. A handheld mic allows you to move around, but remember, the mic should never point at a speaker; this will produce feedback. You should also be conscious of the mic cable when you move. Try not to get tangled in the other musicians' cables, or get the cable wrapped around a speaker or mic stand. These things can and do happen on stage.

10.5: Never cover the mic's element while singing into it!

Avoid covering the mic with your hands when you're singing. Take a close look at figure 10.5. Heather is blocking the mic capsule, which is like putting a sound barrier between her voice and the audience's ears.

Microphones on a Stand

Getting a good sound from a mic on a stand involves two things:

1. Setting the stand in the best position
2. Maintaining a consistent position relative to the mic while singing

Avoiding Vocal Mic Mistakes
Pro Tip by Gino Sigismondi, Shure, Inc.

Vocalists actually make three major mistakes with mic technique. Number 1, they simply don't hold the mic close enough to their mouths. If you are having trouble hearing yourself in the monitor, get closer to the mic! The cheapest, easiest way to get more volume with less feedback is to keep the microphone close to your mouth.

Number 2 is pulling back off the mic too much. The sound engineer needs a consistent level from the singer to keep the vocals mixed properly, and pulling the mic back out to arm's length when you're going

for that high note results in a drastic drop in level. While some variation in mic distance can be helpful, small changes are usually all that's necessary. For reference, moving the mic from 1 inch away from your mouth to 4 inches away from your mouth results in a 12dB drop in level, which is more than half!

Finally, never, ever, ever, cup your hand over the mic grille! I know it looks cool when the rappers do it, but it gives the microphone a hollow, muddy sound and alters the microphone's polar pattern in a way that actually makes feedback more problematic.

10.6: Proper position for addressing a mic on a stand

For singers who don't play an instrument, the stand is just an extension of the microphone. In fact, it makes a pretty good prop. You can hold the mic, or have your hands free to help you express the song.

Figure 10.6 shows a singer addressing the mic on a stand. The stand is at a good height: Her mouth lines up with the capsule without making her tilt her head up or down, two things that can hinder a singer. Note that she's singing into the capsule, but not "eating" the mic.

When you sing and play an instrument, it can be especially hard to maintain mic position. Let's say you're singing lead vocal, and have a solo coming up; it's not uncommon to move your mouth away from the mic capsule at the end of the vocal line in preparation for the solo. The only way to overcome this is to practice singing into a mic while playing. Work on finishing each of your vocal lines without moving out of the mic's range.

Electric Guitar

Electric guitars are designed to plug into amplifiers, which contribute quite a bit to a guitarist's overall sound. (Some even argue that the amp is more important than the guitar itself.) So while you *could* run an electric guitar's signal to a direct box, it's way more common to send the guitar to an amplifier or to a preamp designed specifically for guitar.

Guitarists often plug effects in between the guitar and the amp. Effects may also be plugged into an amp's built-in effects send and return, known as an effects loop—a common feature on modern amps. For now, let's assume the guitar is going straight to the amp. You can:

1. Turn the amp up loud enough that you don't need to mic it
2. Mic the amp
3. Take a direct out from the amp (if available) and send that to the house

MIKING A GUITAR AMPLIFIER

The tried and true method for miking a guitar amplifier onstage is by placing a cardioid dynamic microphone close to the speaker. Ribbon mics have also become popular for this; some large-diaphragm condensers can work too, as long as they can handle the sound pressure levels produced by the amp.

The mic's capsule can either point directly at the speaker (on axis) or slightly away (off axis). You'll have to listen to determine which sounds best.

Close Miking

The mic can be placed so close to the amp that it touches the speaker grille, as shown in Figure 10.7. This will provide the most isolated sound from the amp, since the speaker cabinet itself will block sound from other sources.

10.7: A close-miked guitar amp with the capsule pointing on axis

The mic may sound better when pointed at an angle relative to the speaker. Figure 10.8 shows an off-axis mic position. Listen to both options to decide which is best. Remember, unless you have the amp at very low volume, the sound coming through the P.A. will be blending with the sound from the speaker cabinet. The mic can therefore be used to emphasize a certain aspect of the sound (for example, bring out a little more upper midrange bite or lower midrange thump).

Moving the Mic Away From the Amp

Depending on the amp, the speakers in the cabinet, the volume you're playing at, and the microphone's proximity effect, such ultra-close miking might not produce the best sound. For example, a cabinet with two or four speakers will produce a sound that's hard to capture

10.8: Pointing the mic off axis on the speaker's edge

with one close mic. Try moving the mic back a few inches and see if that's better (Figure 10.9). You may need to put a barrier to the side of the mic to isolate it from other noises onstage.

10.9: A dynamic microphone a few inches from a two-speaker guitar cabinet

Hanging the Mic

Some sound engineers will hang a microphone over the amp. If the mic captures sound at the side, this will offer similar results to pointing the capsule at the speaker. If the mic captures sound at the front, this will be *really* off axis—the mic will be picking up sound from the floor. Some people hang a mic because they don't have (or don't want to set up) a mic stand. It can work, but most engineers consider it to be a last resort. If mic stands are an issue, consider getting a clip-on mount that will stay with the amp.

Phase

When two or more speakers are producing the same sound, the sound waves they create can be out of phase at certain listening positions. This can be desirable, or it can make the sound thin. If the amp sounds good on its own, and the sound going through the P.A. sounds good on its own, but the two sound tinny or thin when combined, phase may be the problem. Try moving the mic. If the mixer has a phase switch, this *may* help, but it may not. Always judge with your ears.

Electric Bass

The electric bass is similar to the electric guitar in that it typically plugs into its own amp on stage. But if the bass is to go through the P.A. system, it usually runs into a direct box first. The balanced output from the direct box goes to a mic input on the mixer; the unbalanced signal goes on to the bass amp or preamp, as shown in Figure 10.10.

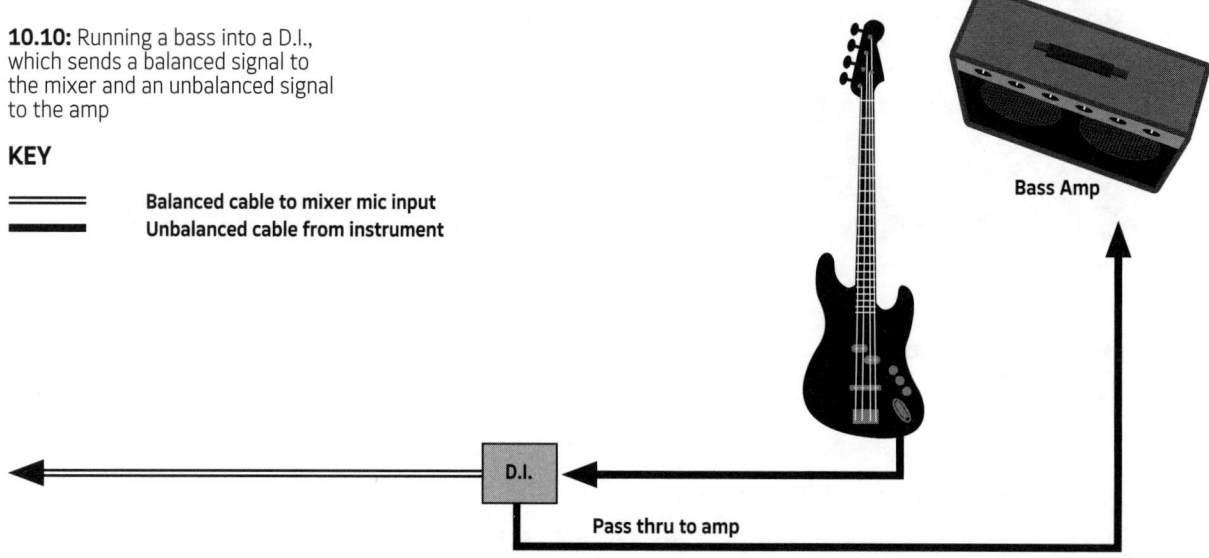

10.10: Running a bass into a D.I., which sends a balanced signal to the mixer and an unbalanced signal to the amp

KEY

Balanced cable to mixer mic input
Unbalanced cable from instrument

Bass Amp

D.I.

Pass thru to amp

Why? Low frequencies are very difficult to control in any room. They're also very difficult to capture with a microphone—especially a close mic. The direct signal will actually send more low end to the house system than a mic on the amp.

If you use a pedalboard, it should go between the bass and the D.I. so that any coloration you add with effects can also be heard in the house.

Direct Feed From a Bass Amp or Preamp

You can also take a direct feed from the bass amp. This gives the player more control over his or her own tone, while giving the sound engineer less control. If the bass player turns up or decides to boost bass, the front-of-house sound will have to be adjusted accordingly.

Because the bass runs direct so often, many bass amps and preamps have features that make it easier to connect to the P.A. mixer. These include XLR balanced outputs (Figure 10.11), speaker mute (which lets the amp act as a preamp), and separate level controls that let the bassist boost an amp's signal onstage without affecting the level at its D.I. output.

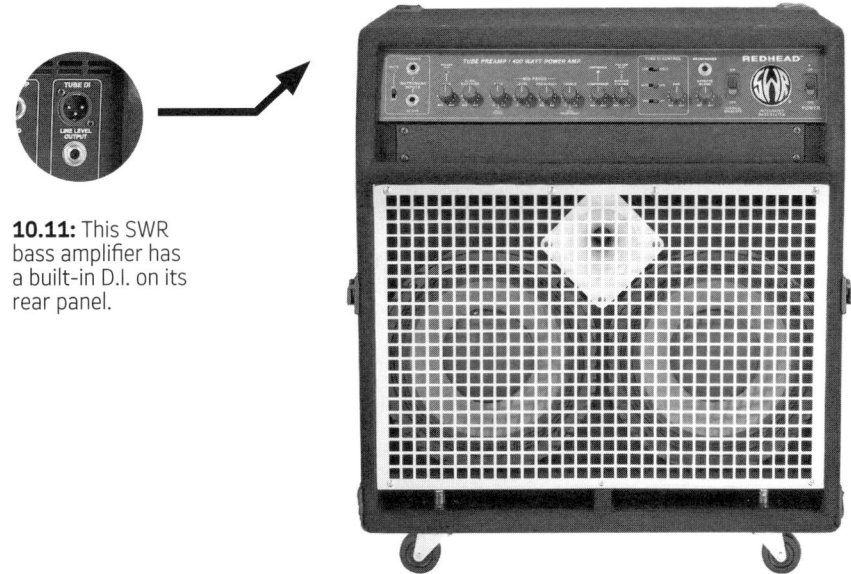

10.11: This SWR bass amplifier has a built-in D.I. on its rear panel.

If you know you're going to be running direct, you might want to bring a direct box of your own. Compact floor-mounted preamps (Figure 10.12) offer the player some control over his or her sound while providing a balanced signal for the house.

10.12: A floor D.I. preamp designed for bass

Acoustic Guitar and Other Acoustic Strings

Acoustic guitar can be challenging to capture effectively onstage. The same holds true for acoustic stringed instruments such as the violin, viola, cello, upright bass, mandolin, ukulele, banjo, and others. We'll use the acoustic guitar to illustrate the general method for capturing these instruments, though we'll add more instrument-specific tips along the way.

Part of the problem is trying to decide which of the many options works best. An acoustic guitar can run exclusively through the P.A., through an amplifier, or both. Either way, the guitar may be captured by a pickup (sometimes by a multi-pickup system), by a microphone (or pair of microphones), or by a combination of mics and pickups.

MIKING AN ACOUSTIC STRINGED INSTRUMENT

Miking any stringed instrument can be a delicate matter. The instrument itself produces a complex sound that changes depending on your listening position. In an ideal setting, you could walk around the instrument, listen for where it sounds best, and place the mic there. In the real world of live performance, you'll find yourself close-miking the instrument most of the time—anywhere from three inches to a foot or so away. You can even place a miniature mic on or inside the instrument, though we'll consider this more as a pickup than as a mic.

Many engineers prefer small-diaphragm condenser mics for acoustic instruments. As long as the mic is directional and is positioned correctly, it should work onstage with minimal bleed and feedback. A solo performer in a good acoustical space might even use a pair of condensers.

A dynamic mic won't capture as much detail as a condenser, but may be more practical or may be the only thing available. Either way, mic position is critical.

Positioning the Microphone

A directional mic (cardioid, supercardioid, or hypercardioid) will offer the best isolation, but it will also focus on a particular part of the sound, instead of capturing the general sound of the instrument. Therefore, mike position is really important. On guitar, there are

Choosing a Mic For Acoustic Instruments
Pro Tip by Gino Sigismondi, Shure, Inc.

Condenser microphones tend to be best for acoustic stringed instruments. The extended high frequency response and better transient response of condenser mics brings out the nuance of acoustic instruments. Pickups are a viable option when you need a quick, convenient way to mic an acoustic instrument. You get great isolation and minimal feedback, but the sound quality is typically inferior to what you'd get with a good microphone.

two common places to point a mic: towards the bridge (Figure 10.13) or towards the sound-hole (10.14). Each will offer a different sound, and the distance between the instrument and mic can also make a difference.

10.13: Pointing a mic towards the bridge gives the sound more body.

10.14: Pointing a mic towards the sound hole can work, but may sound a little boomy.

Many people assume that the soundhole is the best place to point the mic, but this can actually yield a boomy and inconsistent sound; the bridge will offer more body. You can also try angling the mic toward the neck to capture more string detail.

Marking the Mic Position

Once you set up the mic and determine where it sounds best, you should mark the spot so that the player knows where he or she is supposed to stand. Players like to move on stage, and this will affect the sound. You might even work with the player to set up two spots: one for louder playing such as strumming, and a closer spot for lead playing and quieter passages. If you're a player, work the mic and take note of where the instrument sounds best.

Be sure that the mic doesn't interfere with the performance. Even if it sounds better where it is, the musicians' ability to play well should always be the first consideration. If the mic is too close for comfort, move it back and do your best to adjust the sound as needed.

PICKUPS

Acoustic instrument pickups, as mentioned earlier, come in two types: magnetic pickups (which work like electric guitar pickups) and contact pickups (such as the piezo pickup mounted under an acoustic guitar's bridge, or pickups that use sensors to read body vibrations). Also, clip-on or internal microphones are often treated like pickups by engineers.

Some systems combine more than one technology—for example, an acoustic guitar using a bridge-mounted piezo in tandem with a mic sitting inside its soundhole (Figure 10.15).

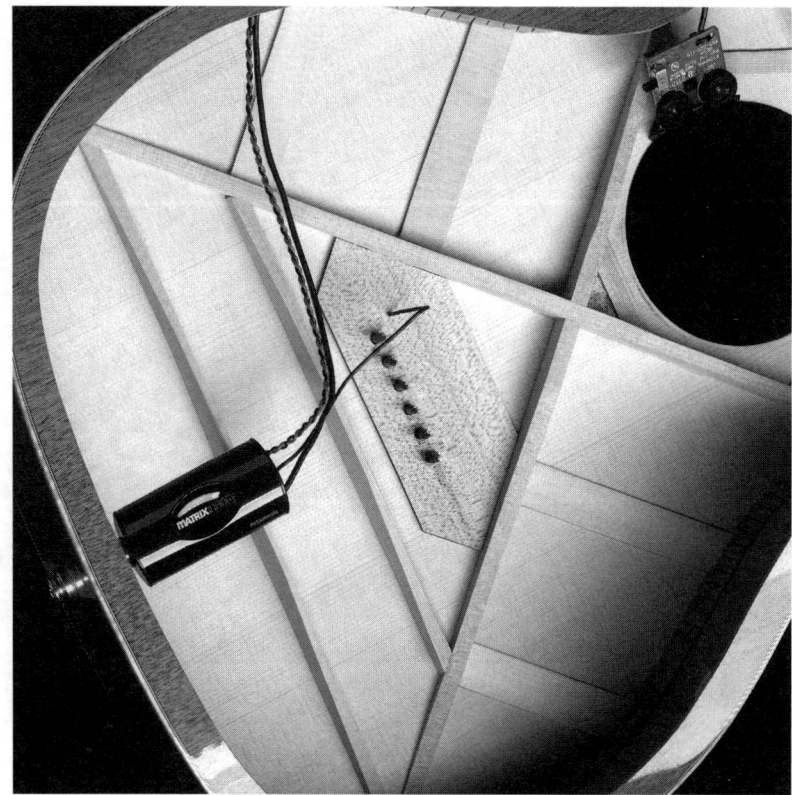

10.15: An interior view of a Fishman pickup system mounted in an acoustic guitar.

Passive and Active Pickups

A passive pickup is simply a device that reads the strings and produces a signal. No other electronics are included with the pickup. A passive pickup can feed an amplifier or a direct box, but it will have a lower output level than an active pickup.

An active pickup uses electronics to boost the signal, and is usually mated to an onboard preamp that includes an equalizer, volume control, and more.

Most acoustic instrument amps and preamps work with both active and passive pickups. You'll often find a switch to help set the amp's input level.

AMPS VS. DIRECT

Unless you're going for a specific sound, you're probably not going to like the way an acoustic instrument sounds through a guitar amp. The guitar amp's speakers will cut off a lot of the treble that give an acoustic instrument its character. Amps specifically designed for acoustic guitar (which also work with other strings) sound much better. In a small venue, such an amp may be all you need.

Still, acoustic stringed instruments are often taken direct through a preamp or direct box. The same techniques used for bass apply here: The instrument can go into the D.I. with a balanced line feeding the house system. If the player wants to use an amp, an unbalanced line can go from the D.I. to the amp's input. You can also take a feed from the amp's line output and send that to the house system.

ACOUSTIC BASS

Like the acoustic guitar, upright bass can be miked or captured with a pickup. Miking a bass in a loud environment can be especially tricky because the instrument itself isn't very loud and doesn't always project a clear tone. In the studio, ribbon or large-diaphragm condenser mics can be effective when positioned a few inches from the front of the instrument, and this might work in a quiet performance setting, as well.

If you're working with the bass alongside amplified instruments, however, a pickup or a body-mounted mic like the one shown in Figure 10.16 is probably a better choice. Some upright players prefer to run a pickup to a bass amp in addition to (or even instead of) going direct to the mixing board through a D.I. box.

10.16: A DPA pickup mounted to the bridge of a string bass.

VIOLIN, BANJO, AND MANDOLIN

Other stringed instruments can also sound good when captured by a single small-diaphragm condenser, though again, a dynamic mic may have to suffice. Many players now use contact pickups or clip-on mics.

If you do use a free-standing microphone, try the following:

Violin, Viola, and Cello

Some engineers like to place the mic above the instrument—often 18 inches or more—with the capsule pointing at the top. That may work for an acoustic ensemble, but on a louder stage, you're probably better off positioning the mic about three to six inches from the instrument, pointed to the side.

Violins using onboard mics and pickups can be treated like acoustic guitars (Figure 10.17). Electric violins—that is, those designed specifically for amplification—might also work through an electric guitar rig.

10.17: L.R. Baggs contact mic mounted to violin bridge

Mandolin and Banjo

Mandolins can be close-miked or use an onboard pickup, either built into the bridge or added on later. Banjos can be close-miked from about three inches away. Pointing the mic toward the center of the body will produce a bassier sound. Pointing it toward the edge will produce a brighter sound. Contact mics and pickups are also available for the banjo.

Miking in Stereo

In the isolated environment of the recording studio, many engineers like to use stereo mics. On stage, this can present a few problems: double the chance of feedback, phase issues, mics picking up bleed from other instruments, etc. Two mics can also get in a player's way more easily than one. One common method of stereo miking is called the X/Y technique, where

Fiddling and Pickups
Fiddler and recording artist Amanda Shires

I have an L.R. Baggs pickup, and have cable from the pickup go either through a preamp and straight into the mixing board or through an amplifier, which I have miked. The only problem: if you're playing with a really loud band, you get a lot of feedback. My friend Tom Hagerman, who plays with the group DeVotchKa, has a little mic that he mounted on the instrument, which is less cumbersome and more sensitive. As far as EQ, etc.: I like to be in control of it as much as I can. I prefer to use a preamp when I'm not using an amplifier and set the knobs myself. My fiddle tends to be a little shrill, so I like to turn down the high end and some of the mid and boost the bass a little bit. But if you work with sound engineers who actually know what they're doing, you can leave it to them. I like to use reverb. I don't use any damping to keep feedback down: I try to get the other band members to turn down.

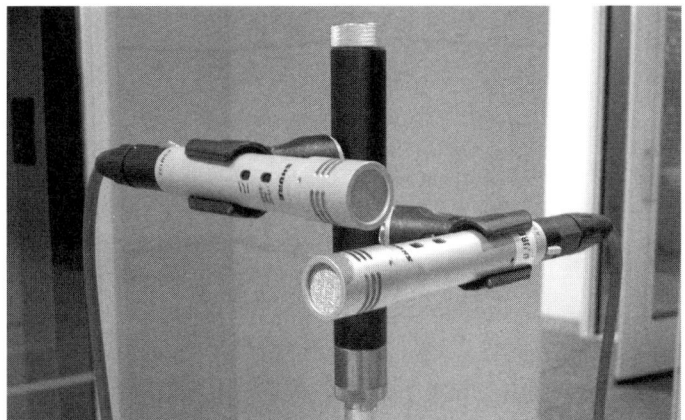

10.18: X/Y stereo miking (IMAGE ©SHURE INC. USED WITH PERMISSION.)

two mics are mounted together in a crisscrossing pattern (Figure 10.18). Keeping the mics close together helps reduce phase problems. This technique is also used for piano and drum overhead miking.

Wind Instruments

There are two schools of thought on miking wind instruments: Point the mic at the bell of the horn, or point it at the body. Wind instruments are pretty loud, so dynamic mics are a common choice. Modern ribbon mics like the Royer 121 are also gaining popularity. A hardy condenser might work as well.

FREE-STANDING MICS

Placing a directional mic close to the bell of a horn or saxophone can offer good isolation. In figure 10.19, a student from the Lagond Music School demonstrates good microphone position with a trumpet, with the bell facing the mic capsule. A similar position can be used for saxophone, though some players opt to place the mic on the pipe instead. Flute should be miked closer to the mouthpiece (Figure 10.20).

As with a vocal mic, position is critical and largely up to the player. A boom stand is

10.19: Trumpets should be miked in front of (but not inside) the bell.

10.20: Flutes should be miked near the mouthpiece. You can use a windscreen to reduce breath noise.

handy when miking the sax because it's easier to point the mic down at the bell. Mics at that level can easily get bumped, though, so the player should look at the mic and be ready to make adjustments.

Figure 10.21 shows bad (left) and good (right) mic positions for the sax.

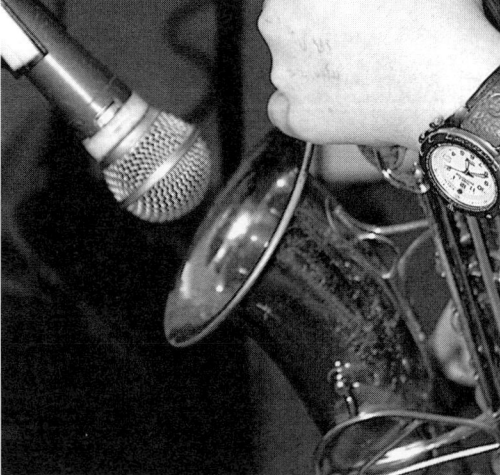

10.21: If a mic is pointing away from the sound source (left), the player should adjust the mic instead of trying to play into it (right).

HANDHELD MICS

Harmonica is the only wind instrument that's miked with a handheld. You can use a vocal mic feeding the P.A., or it can go into a small electric guitar amp using a specially designed microphone such as the Shure DX520, which has an unbalanced 1/4" output. The mic should be held right up against the harmonica (Figure 10.22). To get the famous Chicago blues sound, use a small tube amplifier, and mic the amp through the P.A. if necessary.

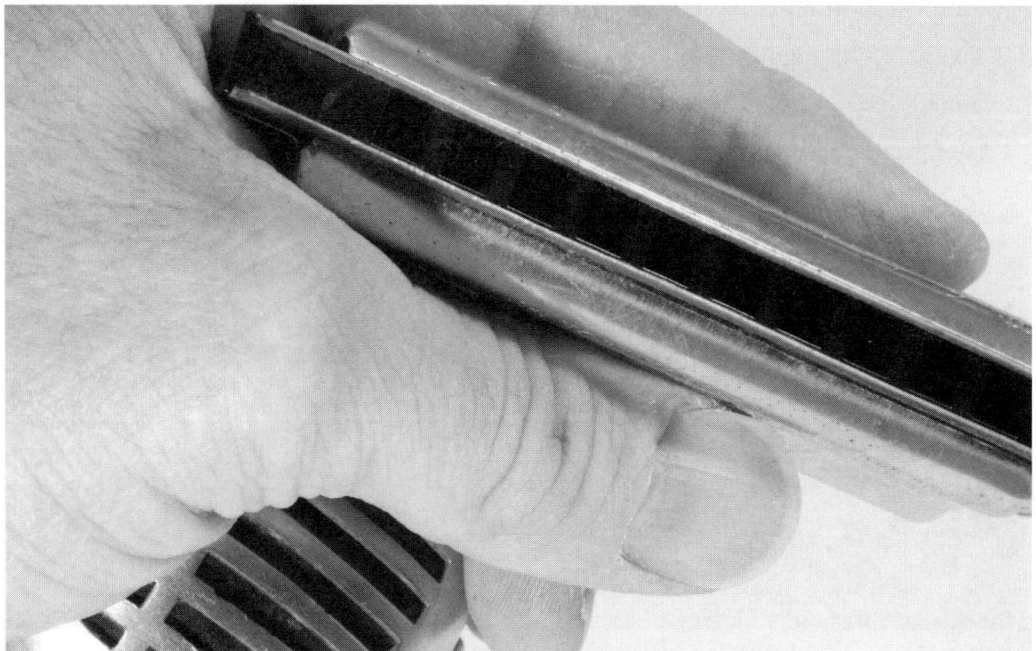

10.22: Harmonica players should hold the microphone right up against the instrument.

CLIP-ON MICS

Horns can also be used with clip-on mics. Figure 10.23 shows a Lagond student with a mic mounted on the bell of his saxophone. Figure 10.24 shows a close-up view of clip-on mics for horn and flute.

10.23: Student with clip-on mic

10.24: Clip-on mics for flute (above) and horn

Piano and Electronic Keyboards

Piano is difficult to capture with a microphone. For one thing, the instrument covers an incredible pitch range. For another, it's physically big. Point a directional mic at one part of the instrument, and you'll miss sound coming from the rest of it.

If you're using one mic, you'll have to make some choices. On an upright piano, you might open the top and place the mic above the treble strings for a good general sound. Professionals usually prefer to use two mics for piano—one for the bass strings, one for the treble. Both mics will pick up the midrange. In a quiet environment, they can be arranged in an X/Y pattern or as a coincident pair (Figure 10.25) a foot or two from the instrument, with the cover off to allow the sound to project.

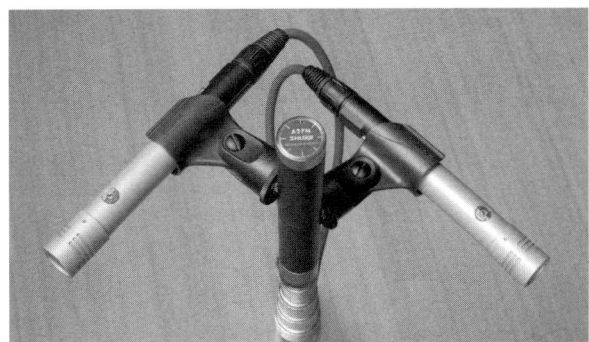

10.25: Two small-diaphragm condenser mics set up in a coincident pair can be effective on piano as well as other sources.
(IMAGE ©SHURE INC. USED WITH PERMISSION.)

In a noisier environment, you can mount the mic to the surface of the piano or try using a boundary mic, which attaches to the surface and acts like the pickups and clip-on mics we discussed in the sections on acoustic strings and winds.

ELECTRONIC PIANO VS. ELECTRIC PIANO

Since real pianos are so hard to mic, you're probably better off using an electronic instrument with digitally sampled piano sounds on any stage that's going to get loud. Such "electronic" pianos are not the same as electric pianos—an instrument that's more akin to the electric guitar, in that it uses amplification to enhance acoustical vibrations. If you're lucky enough to find a working vintage electric piano, you can run it into an amplifier or take it direct. Some players actually like to use tube guitar amps with their electric pianos.

SYNTHESIZERS, SAMPLERS, AND OTHER ELECTRONIC INSTRUMENTS

Electronic keyboards may be the easiest of all instruments to amplify on stage. Most of them have outputs that are designed to run into a mixing board. You may not even need a direct box. They have onboard effects and can store sounds as presets. This lets you set up a whole group of sounds for live performance ahead of time.

Like bass and acoustic strings, you have the option of running direct, using an amplifier, or a combination of the two. If your system includes a number of electronic instruments, you might consider using a submixer onstage. A keyboard amp like the one shown in Figure 10.26 can act as a submixer, combining several channels into one D.I. feed going to the house system.

10.26: This Roland keyboard amp can act as a submixer and send a balanced signal to the main mix.

Mono or Stereo

Most electronic instruments are stereo, but you may not need both channels onstage. For one thing, the house mix may be mono, in which case you're wasting an input by running a stereo feed. You may have to adjust a few things on the keyboard to make a sound work in mono, so it pays to find out ahead of time if you're running in mono or stereo.

Using Multiple Outputs

The sheer variety of sounds you can get from an electronic instrument poses some challenges for the sound mixer. For example, you might use a bright-sounding digital piano on one song and a deep synthesizer bass on another. Unless you account for the difference, you might drive the sound engineer—and the audience—nuts.

If the keyboard has more than one set of outputs, you could assign the heavy synth bass to its own output and feed that to a separate mixer channel from the one(s) you're using for the piano. The sound mixer could then adjust each one independently.

SOLVING THE "HOME" KEYBOARD PROBLEM

Keyboard instruments that are designed for home use may not have the kind of outputs needed to plug into a mixer. For example, you might find a pair of RCA outputs. Or there may be internal speakers and *no* plug for outputs. Rather than trying to mic the speakers (which will probably sound pretty bad), you can use a splitter cable, as described in Chapter 6, to run the headphones feed from the keyboard into two channels of your mixer (Figure 10.27).

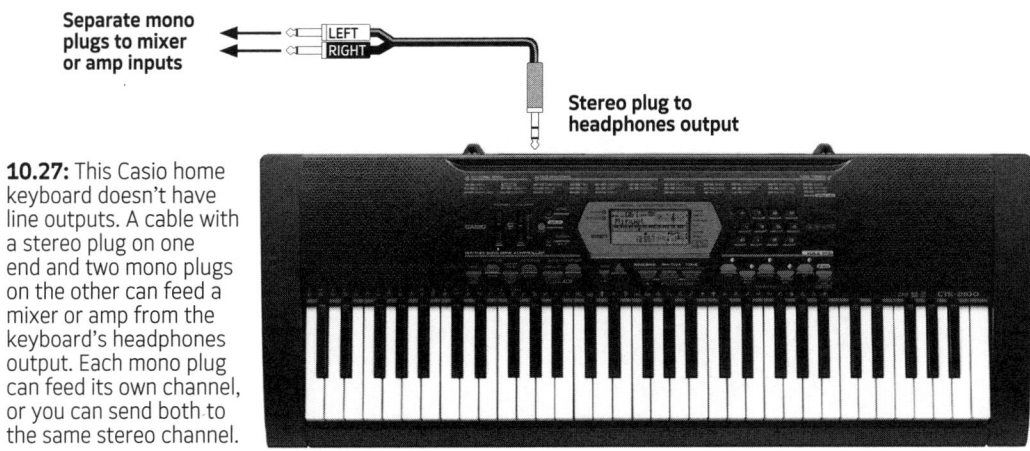

Separate mono plugs to mixer or amp inputs

LEFT
RIGHT

Stereo plug to headphones output

10.27: This Casio home keyboard doesn't have line outputs. A cable with a stereo plug on one end and two mono plugs on the other can feed a mixer or amp from the keyboard's headphones output. Each mono plug can feed its own channel, or you can send both to the same stereo channel.

Drums and Percussion

Miking a drumset is an art in and of itself. The mics need to be close enough to the drums to be isolated, but not get in the player's way. You have to worry about things like phase cancellation and bleed among the mics. Getting a good sound takes practice. One reason why many venues ask performers to use the house drums is so that they don't have to go through the chore of miking a kit over and over.

That said, it is easier to mic drums today than it used to be. Mic kits designed specifically for drums make it simpler to choose the right mic for each drum and cymbal. Many of these kits have mounting hardware to make it easier to position the mics.

MIKING THE DRUM KIT

There are many opinions about what mics are best on which parts of a drum kit. Probably the most common approach is to use one dynamic mic on each drum in the kit, and use condensers overhead for the cymbals. Some more elaborate setups might use two dynamics on the snare (top and bottom), a separate condenser on the hi-hat, or even a second mic on the kick. We'll keep things simpler and use one microphone per drum, plus two overheads.

Placing the Mic

The ideal position of a drum mic depends on the drum and the sound you're going for. You can use conventional stands, but clip mounts such as the one shown in Figure 10.28 are way more convenient. These clips can mount onto the rim itself, so you don't have to worry about carrying a carload of mic stands. They're also easy to position at the far edge of the drumhead, which keeps the mic out of the player's way.

10.28: A clip attached to the rim of this snare drum is being used to mount a Beyerdynamic mic.

A short mic stand—or a boom on a conventional stand—can be used to position the mic on the kick drum. Overheads will require one or two tall and sturdy booms. You can also hang them from the ceiling, but it will be harder to position the mics with any precision or keep them in place if you have them dangling on their cables.

Snare

A single dynamic close to the top head at the rim's outer edge works well (Figure 10.29). Some drummers like to move the mic to the side, but this offers less isolation. Players who use a lot of dynamics might sound good with a mic that picks up some hi-hat and tom.

10.29: A stand-mounted Shure SM57 dynamic mic points at the edge of the snare head. (IMAGE ©SHURE INC. USED WITH PERMISSION.)

Bass Drum

Damp the inside of the drum with a pillow, and mic the drum from the front. Place the mic inside the shell, with the element pointing toward the beater head (Figure 10.30). If your front head doesn't have a hole for the mic, you may need to remove the head. While you could mic the bass drum with any dynamic, it's better to use a mic that's tuned to the instrument's low frequencies.

10.30: A Shure B52 kick drum mic positioned inside the front head of a kick drum.
(IMAGE LEFT ©SHURE INC. USED WITH PERMISSION.)

10.31: A matched pair of small-diaphragm condenser mics are being used as overheads (top) in an X/Y configuration. Note that each rack tom (center) has its own clip-mounted dynamic mic.

Cymbals

Crash and ride cymbals are rarely miked individually. Instead, *overheads*—mics positioned above the drum kit—are used to capture all of the cymbals. Overheads will pick up the overall sound of the kit as well, and you can get very good drum sounds using no other mics. (Tip: If you want to keep the overheads focused on those sizzly cymbals, use an equalizer's high-pass filter to keep the low and midrange frequencies produced by the kick drum, toms, and snare out of the mix.)

If you have a single overhead mic, place it dead center about a foot over the drummer's head. If you have a pair of overhead mics (ideally, they should be the same make and model), you can position them in an X/Y pattern dead center (Figure 10.31), or spread them out to the left and right of the drum kit.

As for that workhorse known as the hi-hat, some drummers and sound engineers like to mike it individually using a small-diaphragm condenser or a dynamic that has a strong high-frequency response. Others leave it unmiked and let the snare and overhead mics capture its sound.

Tom-Toms

For rack toms with heads on both the top and bottom, position the mic very close to the top head at the outer edge of the rim. You may be able to pick up two rack toms that are close together using a single mic mounted in between the two drums (Figure 10.32).

If the drum has no bottom head, you might try mounting the mic inside the tom, though this can produce some tricky resonances. For floor tom, you may want to use a mic with a stronger low-frequency response to bring out the drum's low midrange.

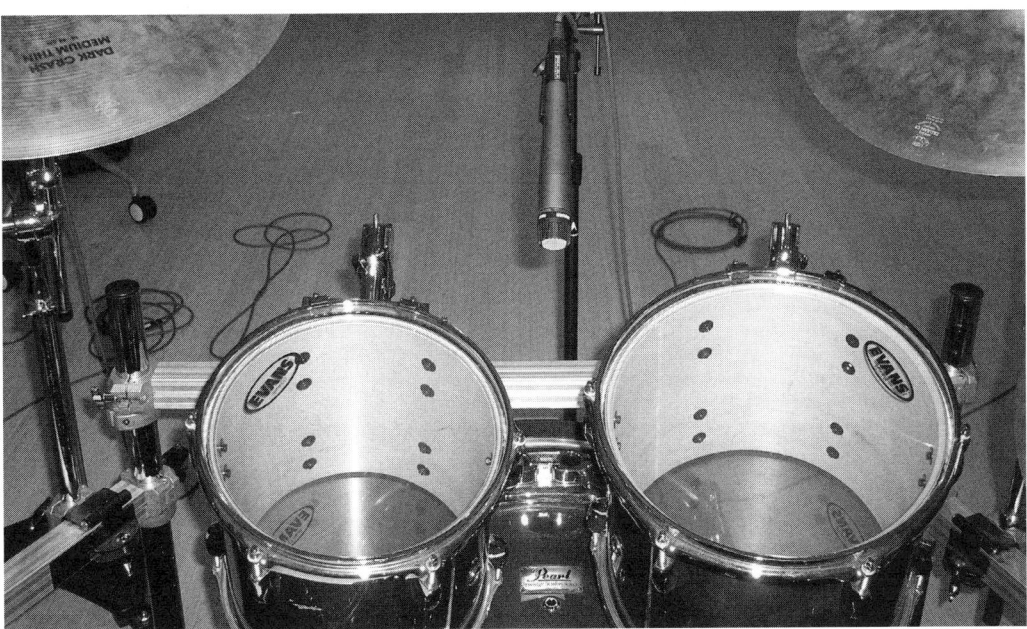

10.32: A single dynamic mic in front of the kit can capture both rack toms. (IMAGE ©SHURE INC. USED WITH PERMISSION.)

The Complete Kit

As you can see, it takes a lot of microphones to cover a drum kit. If you plan to mic every drum and have a pair of overheads, you'll need at least seven mics for a relatively small five-piece kit.

Assembling a good mic collection can get expensive. Professional engineers often like to mix and match different models of mics, often by different companies. But you can save money—and eliminate some of the hassle of choosing which mics to use on which drums—by buying a set specifically designed for drum kits.

10.33: This collection of Peavey mics is designed to close-mic every drum in a five-piece kit (kick, snare, and three toms). Such sets—which are also available from other manufacturers—have become especially popular for working musicians.

Shure, Audix, Sennheiser, and Peavey are among the companies offering mic collections for drums. Their sets will often have mics specifically tailored for kick, toms, and snare. They may even have mounting hardware (Figure 10.33).

If you don't have enough mics or mixer inputs to mic up every drum in your kit, don't despair. You can get a good sound with just a few well-placed mics. We've already discussed how overheads can capture the whole kit. You might add individual mics for kick and snare to give them a little more oomph. If you only have two mics and don't like the way they sound as overheads, try placing one on each side of the kit to capture its overall sound. If the band doesn't play too loudly, you might be able to use one mic on the kick drum for some extra punch and leave the rest of the drums unamplified.

No matter what setup you use, however, be conscious of phase issues when using multiple mics. Don't be afraid to flip the phase switch on any given mic channel and see what sounds best. Sometimes, just moving a mic an inch or two can solve the problem.

PERCUSSION

Percussion instruments are so varied that it's almost impossible to generalize about miking them. For hand drums with skins, you can place mics close to the head(s) to capture the slap of the player's attack, or move them a foot or so away to capture more of the overall tone. If the instrument has a lot of low end, you might need two mics: a close one for the attack and a slightly farther one for the resonance.

With shakers, bells, tambourines, and other small hand percussion, the player will be able to move the instrument to the best position for the part he or she's playing. A good small-diaphragm condenser will capture a lot of the detail, but a dynamic vocal mic can work pretty well too. Drumset players who double on hand percussion can generally use the overheads to capture their sound, though they may need to stand up to do so. (It's better to do that than to move one of the drum mics.)

ELECTRONIC DRUMS

Electronic drums can include sounds generated by programmable machines and those triggered by a player in real time. In terms of audio routing, there's very little difference between the two, since all electronic drums must be amplified to be heard. In that sense, they're a lot like the keyboards we discussed earlier in this chapter. Here, we'll see how a drummer might use electronics as part of—or instead of—acoustic drums and percussion.

Playing Electronic Drums

An electronic drum setup has two main parts. The first is the unit commonly known as "the brain." The brain can store hundreds of drum, cymbal, percussion, and other sounds in its memory. It also includes audio controls capable of adjusting and mixing these sounds, as well as all the connections needed to send them to an amp or mixer.

The second part of an electronic drumset consists of the various pads and sensors a drummer can use to trigger the sounds stored in the brain. Figure 10.34 shows a large kit with individual drum and cymbal pads, as well as a brain mounted near the drummer.

When a player strikes a pad, it sends a signal to an input in the brain, which is programmed to generate the appropriate sound for that pad. You can set the pads to trigger natural drum sounds or use them to create sounds that would be impossible on a real drumset. You can store a set of sounds as a kit, and most brains can hold hundreds of kits.

In addition to being able to generate so many sounds, electronic drums don't need to be miked, so they can be easier to set up. The drummer can even control his or her own

10.34: The Yamaha DTX 550 electronic drum kit has five drum pads plus hi-hat, crash, and ride cymbal pads. The box on the upper left is the brain.

balance with the brain's controls and send that mix to the house system. However, some sound engineers may ask for individual outputs for each drum sound instead. This gives the people doing front-of-house sound the ability to set the balances the same way they would with a miked kit.

10.35: The Roland HandSonic combines pads and a programmable brain in a single unit.

Not all electronic drumsets are complete kits. Compact pads can be played with hands or sticks. Often, these will have the pads and the brain in a single unit (Figure 10.35). You can also use acoustic drums to play electronic drum sounds by mounting triggers on the drums.

Besides generating sounds of their own, most electronic drums have MIDI connections that allow them to trigger sounds in compatible electronic devices, including keyboard synthesizers and samplers, computers, and more.

Prerecorded Tracks

The best way to handle prerecorded tracks depends on both the device playing the tracks and the way they'll be used in your performance. If you want to use a playback device on stage like an iPod, CD player, or even an old-school tape recorder, then you should probably use the method described for "home" keyboards above.

If the tracks are on a computer, you can run the computer interface into the house system using the method described for electronic instruments.

If the music is part of a DJ's performance, the DJ mixer's output can feed the house system as a submix, as outlined earlier in this chapter.

If the tracks are going to be part of a song that also includes live performance, it's essential that the band members get the track in their monitors, so be sure to tell the sound engineer what you're doing.

No matter what, avoid blasting the prerecorded music through a pair of cheap speakers, such as those on a boom box. It will sound distorted and will bring the whole performance down a peg.

Miking Ensembles

Whether it's a choir, horn section, orchestra, or big band, there may be times when you want to capture a full group sound, rather than isolate individual instruments.

Mics can be positioned above or in front of the performers. Often only one or two mics are needed to capture the section.

SPACED PAIR

Try using a pair of cardioid microphones, placed on stands at about head height (six feet) facing the performers (Figure 10.36). Stereo mics can be placed using what's known as the "3-to-1" rule: The distance between the two mics should be three times larger than the distance between the mic and the source. So if the mics are two feet in front of the ensemble, they should be at least six feet apart. You can go a little wider, which may bring out more low end in the sound. This setup can also work well for live recording, either on its own or mixed with individual stage mics.

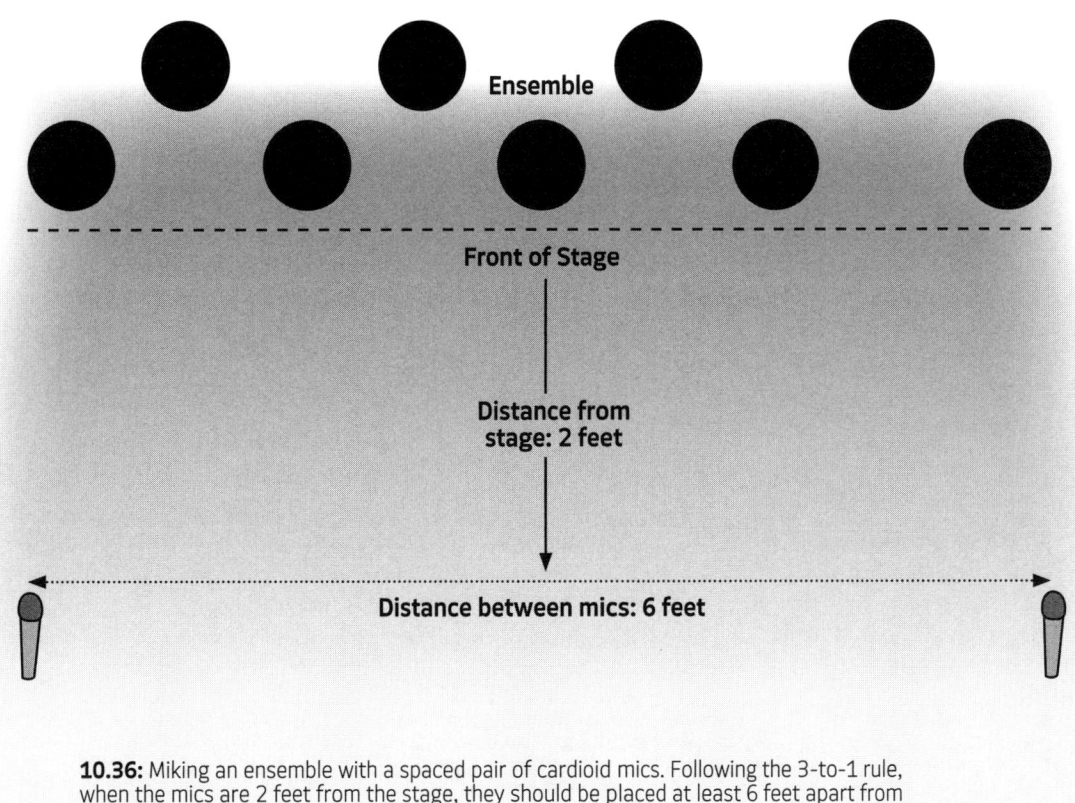

Ensemble

Front of Stage

Distance from stage: 2 feet

Distance between mics: 6 feet

10.36: Miking an ensemble with a spaced pair of cardioid mics. Following the 3-to-1 rule, when the mics are 2 feet from the stage, they should be placed at least 6 feet apart from each other.

Next Stages

Now that you know the basics of capturing the sounds of instruments and voices onstage, it's time to start putting them together to create a mix for your audience.

Chapter 11

SETTING UP THE MIXER

n the previous chapter, we offered a brief and basic overview of how to get sound from the stage to the audience's ears. Now it's time to manage all the microphones and direct feeds coming to the mixing console.

This part of the job is less about technique than it is about being organized and orderly. For now, let's not worry about whether you've got your drum set miked up perfectly; instead, let's focus on the task of connecting your mics and instruments and keeping track of how everything is going to and from the stage.

Plugging In

Hopefully, by now you know that mixers have connections called inputs, and that some inputs are designed for microphones, while others are designed to accept the stronger line-level signals put out by preamps and other electronic gear.

Most of the signals coming from the stage will be running to the balanced microphone-level inputs on your mixer. These XLR inputs can be fed by both microphones (*duh!*) and by direct boxes (not so *duh!*). Unless the mixer is very close to the instruments, use a direct box (D.I.) instead of running an unbalanced line directly from an instrument to a mixer input—even if the mixer has "instrument" inputs.

One of your goals should always be to have a strong signal with as little noise—hiss, hum, or buzz—as possible. In sound engineer-speak, you're looking for a good *signal-to-noise ratio*, sometimes written as S/N. (Quick aside: You'll see S/N statistics for all kinds of electronic devices, and they give you an idea of how quiet or noisy they are, but these numbers aren't as important as things like impedance and level.) If you have a preamp or electronic instrument with balanced line-level outputs, you have the option of running those directly to the mixer's line inputs, or using a direct box.

If the mixer is close to the stage, you might just plug every input directly from the source into the individual mixer channels. If the mixer is positioned away from the stage, use a snake to consolidate the connections. We discussed a lot of this stuff in Chapters 9 and 10.

Before you plug anything in, make sure all amps and/or powered monitors are turned off and no signal is coming through the board. (Most mixers are equipped with input level meters that let you see signal coming into each channel, even when you don't hear it.)

Zeroing the Board

Before setting up a mix, most engineers like to start with a blank slate. Getting the mixer to its default settings is known as *zeroing the board*. On an analog mixer, that means turning every input level to zero, turning down the faders, setting every EQ to its "off" or center

Zeroing In On Your Mix
Pro Tip by Phillip Jordan, Producer/Engineer, DTLA.tv

The purpose of zeroing a board is to give the console a fresh and familiar starting point by returning all its settings to a zero gain and centered position. It's both a courtesy for the engineer that follows you and a good practice after each band's set and before you begin soundcheck. If you're sharing the board with another act, it's important to note every setting for both you and the other performers. If you have a digital camera, use it to take pictures! Or create a template for the board on paper and mark the settings.

When zeroing the board, I usually leave all aux (monitor and effects) masters at unity, but return all group, main, and channel faders to zero. Before soundcheck, I mute all the channels. (On some boards you will be able to include every channel in one mute group.) But if you're running house music from an iPod or DJ, you have to remember to leave those channels active while you work. Here's what I do when I'm working on an analog mixing board:

FOR EACH CHANNEL:
- Return all gain pots to zero
- Disengage phantom power on all channels
- Disengage any hi-pass filters

- Return EQ gains to zero for each band
- Return aux sends to zero
- Center all pan and balance pots
- Bring all channel faders down to -∞

FOR THE MASTER BUS:
- Set all master faders to -∞
- Return all send masters to unity 0dB
- Return all group masters to unity 0dB
- Center all group pans

If you're using a digital board, it's a good idea to carry some kind of storage device (such as a USB flash drive) that you can use to save your settings. Digital boards can store their settings in a preset called a scene (see Chapter 2). Often, they'll have a scene called "Recall 0," which returns the board to an entirely zeroed position. On smaller digital boards, you may still need to physically zero gain pots, hi-pass filters, and phantom power. But with a couple of easy button pushes, you'll be able to zero the rest of the board. Check the owner's manual. And if you're sharing the mixer with others or working on somone else's board, make sure you don't save over their settings!

11.1: A zeroed analog mixer with all controls set to their default positions

position, setting pans to center, etc., as shown in Figure 11.1.

With a digital mixer, you can often zero the board automatically, simply by calling up a new preset. However, some controls—such as input gain knobs—may have to be set manually as well.

CHOOSING CHANNELS

Does it matter which instrument or mic plugs into a given mixer channel? No. Wait....yes. Uhhhh—it depends. Let me explain.

Any microphone can work on any mixer channel with a mic input. If you're using a condenser mic, it will need phantom power to work. As long as that's available, you're fine. A line signal is even easier to connect. It can run to a line input *or* to a mic input, as long as the gain is set to prevent the line output from overloading the mixer.

That said, your mixer might throw some variables your way. One small mixer might offer more bands of EQ on some channels than on others. Another might have inserts, but only on a few of the input channels. A third might have mono and stereo channels—the latter can be handy for taking signal from a keyboard or computer, but they aren't very effective if you're using mono microphones.

So start by studying your mixer and deciding what channels are best suited to which tasks. You might decide that the vocals need to be on the channels with the inserts so you can plug in a compressor. The bass may need the EQ in case the room can't handle the low end. The digital piano might take the stereo inputs, while the analog synth and the drum machine might each take one mono input.

CONNECTING EFFECTS

Mixers can connect to effects in two ways: using the master sends and returns, which are available for all channels, and using individual channel inserts.

Connecting Send and Return Effects

Effects like reverb and delay are normally connected to master sends and returns. This allows them to be applied to multiple channels while allowing the mixer to blend them for each channel, as well as control their overall mix. Players often want to hear effects like reverb in their monitor mixes. On some mixers, you can route the output of one aux bus to another or use a group bus for the monitor. On others, you might have to manually plug the effects into the monitor bus. Figure 11.2 shows routing where the effect feeds both the house and the monitor buses.

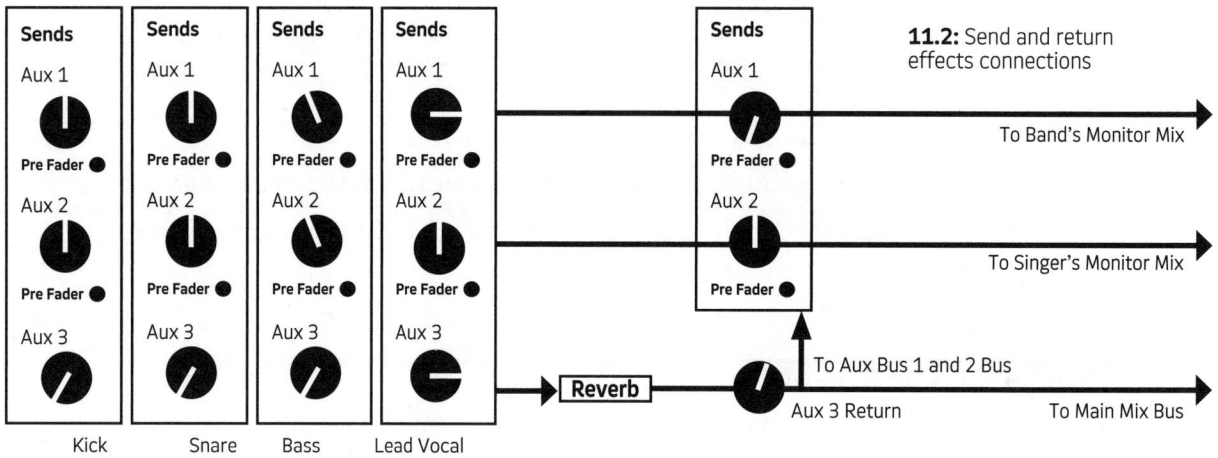

11.2: Send and return effects connections

To connect a send and return effect:

    1. Choose which bus will be connected to the effect. For example, you might use auxiliary bus 4 to connect to a reverb.

    2. Connect the bus's master send to the effect input

    3. Connect the effect output to the bus's master return

The effect and the bus may be mono or stereo. If you're running a stereo effect into a mono bus, consult the effect manual to see which channel to use for a mono connection.

Connecting Insert Effects

Effects like noise gates, compressors, and EQs are often used on individual channel inserts. Inserts are almost always mono signals that use a splitter cable, with a stereo connector on one end and two mono connectors on the other. The stereo end plugs into the mixer channel insert jack. One side of the mono cable feeds the effect's input; the other takes signal from the effect's output and returns it to the mixer (Figure 11.3).

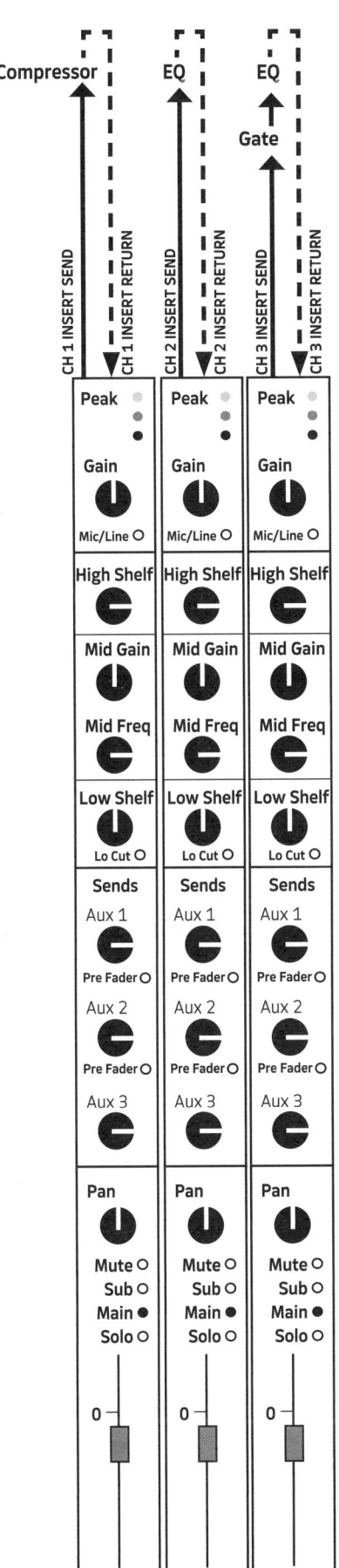

11.3: Insert effects connections

Connecting a Patch Bay

Rather than individually connecting each outboard processor to each channel, you can plug both the processors and the mixer into a patch bay. Then, you can use patch cords to connect any mixer channel to any processor connected to the bay.

LABELING CONNECTIONS

Want to save yourself a lot of hassle? Label *every* connection you have. You can number them, or you can indicate what they're being used for, but there should be a way of identifying individual cables and plugs when they're all massed together.

Since cables using XLR connectors have different connectors on each end—female at the input, where the mic plugs in, and male at the output, which plugs into the mixer—it's useful to put a piece of tape on the input side (Figure 11.4). This makes it easier to get the right side of the cable to the right spot without having to look closely at the connectors.

11.4: Labeling cables

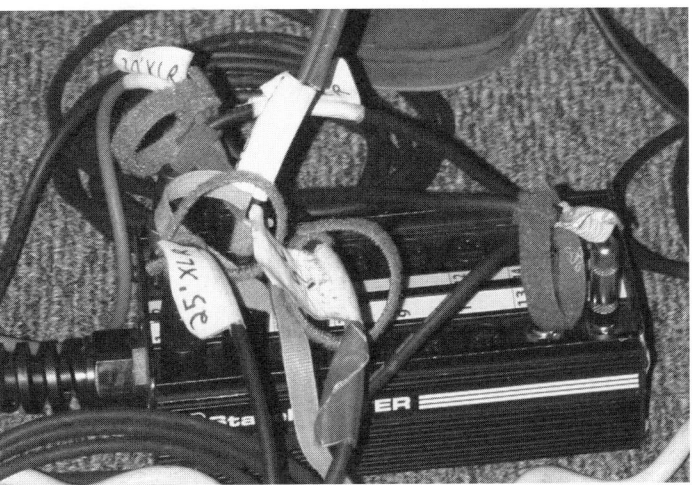

Color Schemes

When a number of cables are combined in a snake, the individual channels are often color-coded. If you use the color spectrum as a guide (red, orange, yellow, green, blue, indigo, violet), you can consistently use red for channel 1, orange for 2, yellow for 3, green for 4, and so on, with violet being channel 7.

Labeling the Snake

If you're using a snake onstage, label its inputs for the corresponding mixer channels. Sometimes, it's pretty obvious: If you have a 16-channel snake feeding a 16-channel mixer, the channels should line up. But if you have, say, two eight-channel snakes, you should indicate which one is for 1-8 and which is for 9-16. If possible, label the snake input with the source; for example, Input 1: Vocal Mic, Input 2: Acoustic D.I., etc.

Labeling Input Channels

As you make your connections, place a thin strip of tape at the base of the mixer, below the faders (Figure 11.5). You may even see a special area, called a *label strip*, marked for this purpose. Use tape that's not too sticky, like 3M 3051 Low Tack Paper Tape, which comes in white. With a thick magic marker, label each channel; try to write as neatly as possible.

11.5: Labeling the channels

Labeling Outputs

Just as you indicated the sources coming into the mixer, it also pays to label the outputs. There might not be a good spot on the mixer itself to do this, but you can keep a sheet of paper next to the board or taped to its side that shows how everything is connected (Figure 11.6). This isn't so necessary if you're running the mixer to a single destination, but it can be pretty helpful if you've got multiple auxiliary sends going to various monitor mixes and effects.

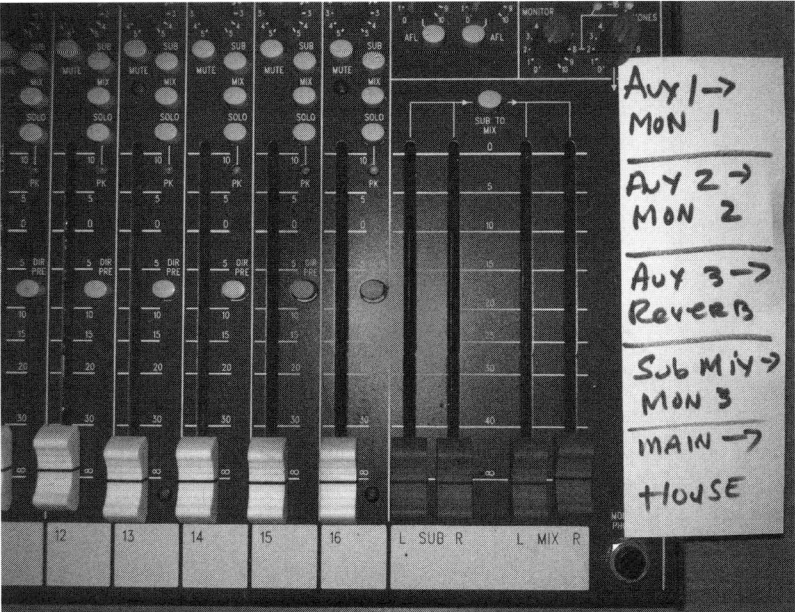

11.6: A note card showing the outputs on a mixer

If a channel is connected to a signal processor at its insert point (see above), you should note that too. While you're at it, stick a note on the signal processor as well. Let's say you're using channel 1 for vocals and have a noise gate at the insert. Show that on the label, and then put a note on the gate showing that it's plugged into channel 1. The more gear you have connected, the more useful these notes become.

SECURING CABLES

Whether you're running individual lines to the mixer or using a snake, you need to keep the cables neat and under control. You don't want to trip over them or pull them out by accident. Cables should be clamped or tied to their mic stands. Cables running to and from amplifiers can be secured by threading through the amp's handle (Figure 11.7). You should

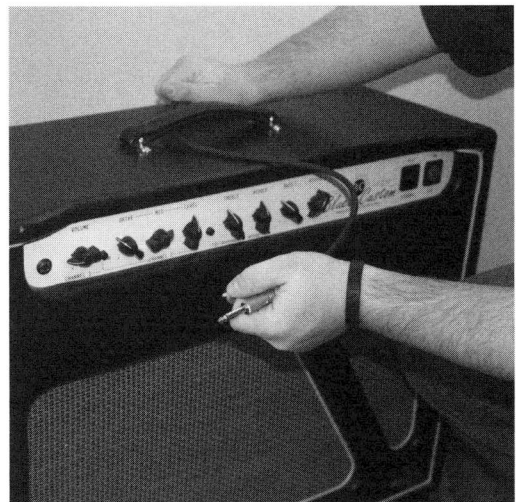 

11.7: Threading a cable through an amp handle (left) or tying it to a mic stand (right) will keep it secure.

11.8: A Lagond School crew member lays cable neatly in front of the stage. **11.9:** A carpet runner can hide and protect cables while making them less of a tripping hazard for the audience.

secure all long cable runs to the floor. Duct tape is very useful for this, both onstage and on the floor between the stage and the mixer. Just be sure to leave enough slack to move the cable ends and make connections. By combining all of the cables into a single path, you can reduce the chances that they'll get unplugged. Figure 11.8 shows cables being laid down in a neat path. Depending on the venue, this could be along the wall or down the middle. Either way, it's a good idea to cover the cable with a rug, a runner (Figure 11.9), or a plastic cable guard. This way, the audience is less likely to step on—or trip over—your cables.

Soundcheck

Once the board is zeroed and everything is plugged in, you're ready to set levels and start your soundcheck. This will involve a number of steps:

1. Checking that all inputs are working
2. Setting the proper level for each input
3. Checking that the house system is receiving signal
4. Checking that the monitor system is receiving signal
5. Checking effects
6. Adjusting the sound of individual channels
7. Checking the ensemble
8. Setting up a house mix
9. Setting up a monitor mix

CHECKING INPUTS

If the channel can switch between mic and line inputs, make sure that it's set correctly for the source. If a microphone plugged into the channel requires phantom power, make sure it's getting that. (But NEVER turn phantom power on with the mixer feeding the house speakers—it'll create a loud pop!)

Next, have one of the musicians, or an assistant, make some noise by speaking into (or gently tapping) the mic, playing the instrument, etc. If the mic has an on/off switch, make sure it's turned on. If the source has a level control—such as the master volume on an electronic keyboard or on an acoustic guitar preamp—make sure it's turned all the way up.

If you're setting up the system before the musicians arrive, you can just do a simple level check. If the musicians are available, you may want to combine this step with the next one and set basic levels. Solo each channel, and listen on headphones to make sure that the sound is coming in clearly with no problems like crackling or ground hum. Don't just rely on the meters. A channel that's producing hum, for example, will fire the meters.

Working with Wireless Mics

If a wireless microphone or instrument is feeding the input:

1. Turn the transmitter and receiver on with the channel muted.
2. Use the meters on the transmitter to make sure that the transmitter is receiving signal from the mic or instrument.
3. Check the meters on the receiver to make sure it's receiving signal from the correct transmitter.
4. Check the channel input to make sure it's getting signal from the receiver.

SETTING LEVELS

Signal coming into a mixer channel should be at a strong level but not high enough to cause clipping. The idea is to have enough level that the faders offer a strong output without having to push them to their maximum.

If the performer is both playing and singing, check the vocal and instrumental feeds both individually and together (that is, playing and singing at the same time). If the performer is using more than one instrument (for example, switching guitars between songs), check every single one of them. If the levels vary among the instruments, make note of it.

For drum kit, first set the input level for each individual drum. Then have the player use the whole kit. Check for phase problems when all the mics are active.

Input Gain

The *input gain* control is the first control the signal will see when it comes into the channel. This is the key to setting levels. If a source signal has low gain (which means it sounds quiet), turn up the input gain control so that the mixer can balance it effectively with the other channels. Otherwise, the sound will get lost, no matter how high you set the channel's output control.

On the other hand, if the source is loud, you must turn down the gain so that the channel doesn't sound distorted. When the gain is adjusted correctly, you can use the faders to set the balance between all the channels. If the gain control is all the way down and the sound is still distorted, reduce the level coming from the source. Use a switch called a *pad* to bring the level down. Some microphones have them. If an instrument or preamp feeding a D.I. is causing the overload, you may need to turn down the instrument's volume or switch the mixer channel from mic to line level.

Read the Meter

An audio mixer usually has a light (or set of lights) next to each gain control (Figure 11.10). Typically, these gain meters use three colors—green, yellow, and red—to indicate the level of the signal coming into the channel. Green means gain is low but signal is present. There's no chance of distortion, but the signal is so quiet that it will need to be boosted. Yellow means it's getting higher but is still safe. Red means the signal is overloading.

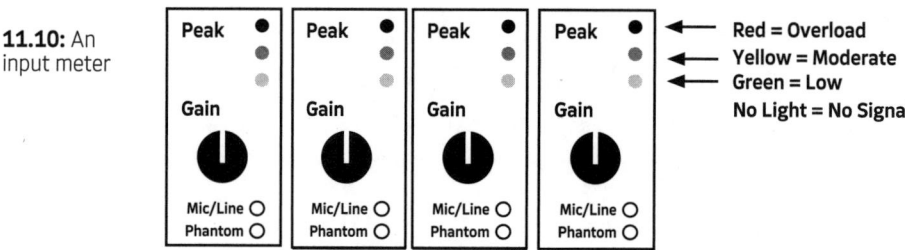

11.10: An input meter

On an analog mixer, it's okay to hit the red occasionally, but if red is showing all the time, the sound is going be distorted. Yet if the light stays green even when the source is at its loudest, you've got the input gain set too low. Why is this bad? If the gain is too low, you may have to overboost the channel with the fader just to be heard. This can lead to noise and other problems.

The ideal setting is when the light is green most of the time, pushes into yellow on the louder parts, and occasionally reaches red on the very loudest peaks. (On a digital mixer, it's better to avoid overloads at all times. Digital clipping sounds especially harsh.) Once you have signal coming into the channel at the correct level, you can use the channel's controls to get it to sound great.

Tips for Setting Individual Levels

The most effective way to set an input level is to have the player or singer run through some part of their performance at the volume they intend to use during the actual show. You should ask them to cover their dynamic range by playing at their normal level, then playing at what they think will be their loudest and quietest levels. Make sure the loudest levels don't overdrive the mixer input. If you're miking an amp, also be aware that players might turn up during the show, and be ready to account for it.

Here's some advice: Always assume that "soundcheck" loudest is going to be a little tame compared to "show" loudest. When the adrenaline is flowing, everything gets louder. You can set the level a little on the low side to give yourself some headroom when the performance energy kicks in.

Check levels one player and one channel at a time (Figure 11.11), but also double check each channel when the whole group plays together during soundcheck.

11.11: A Lagond School house mixer adjusts an input channel during soundcheck.

Tips for Performers

If you've read the section above, you already know that soundcheck volume can be deceiving. Try to be as accurate as possible when you're asked for a level check. And if you're using an amp or a pedalboard, try to keep the settings where you had them during soundcheck. Don't turn up in the middle of the show and expect the mix to sound good!

Checking Sources That Aren't Going Through the Mixer

Onstage amps and drum kits may not need to go through the house mix, but you still need to include them in the soundcheck. Get the players to turn their amps' master volumes to zero, turn the controls on their instruments all the way up, and then slowly turn up the amps' volume controls. When the level seems right, have them mark the spot with a piece of tape, and take note of it yourself.

Check any devices that might boost volume, such as a guitar player's distortion box, an amp's lead channel, etc. Make sure that the boost isn't so high that it throws the mix off. Mark settings for all boosting devices as well.

CHECKING THE HOUSE SYSTEM

With everything turned down on the mixer, turn on the power amps (or powered monitors) and send a signal to the house. If there's no audience in the venue, you might try plugging in an audio test tone generator like the one in Figure 11.12 into an input channel or effects return.

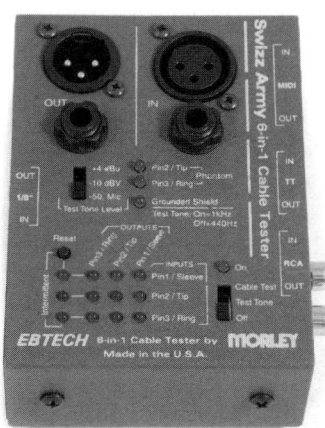

11.12: This compact test tone generator also works as a cable tester.

If you don't have a test tone generator (or if the audience is already in the room), you can feed some house music from an iPod or other portable player into the mixer. Start with the faders on zero and bring them up slowly, then slowly bring up the master volume going to the power amps. Once you've established that the speakers are working, you can bring the levels down again, or continue feeding the house music to the speakers while monitoring the input channels with headphones.

CHECKING EFFECTS

Use the same test method to make sure all effects are receiving signal and sending it back to the mixer at an appropriate level. If the effect has input and output controls, adjust these appropriately. A couple of notes:

1. Effects connected to the send and return are usually set to 100% effect (wet), so that you can control the blend more effectively with the mixer.

2. When effects are connected to channel inserts, you must control the wet/dry mix with the effect. Compressors, gates, and EQs usually have no mix control (unless they're part of a digital multi-effects processor). But if you're using something like a reverb or delay on an individual channel insert, you must set the blend so that the original signal comes through at the desired level.

CHECKING THE MONITOR SYSTEM

If you're using a separate monitor system, run the same check you used for the house speakers into each monitor channel. For example, if you're using three aux buses to create three separate monitor feeds, plug the test signal into a channel and slowly bring up aux 1 for the first monitor to confirm that it's working. Then bring up aux 2, followed by aux 3. Remember, the aux knob must be turned up at both the individual channel and at the master aux send.

Checking Wireless Monitors

Check wireless monitors in two places: where the aux buses plug into transmitters, and where the receiver plugs into the earpieces.

1. Use the meters on each transmitter to make sure it's receiving signal from the mixer.
2. Check the meters on the receiver to make sure it's receiving signal from the correct transmitter.
3. Listen to the headset plugged into the receiver to check signal.

GETTING A BASIC SOUND FOR EACH CHANNEL

Soundcheck is your first opportunity to adjust the sound of each channel and start your monitor mix. Bring each channel up in the house mix to a reasonable level. Check for feedback. If you get some, make sure the mic is not facing the speaker and adjust accordingly.

Using EQ

This is a subjective area, but I advise you to go lightly on the EQ unless you really feel that the sound is getting better when you tweak it. Using EQ to boost a specific frequency can actually increase the risk of feedback. Always A/B your settings by switching the EQ on and off and comparing the sound. One thing you can do, however, is use a hi-pass filter (also known as a lo-cut) to eliminate low frequencies from all vocal mics and treble instruments (Figure 11.13). This will increase headroom in the mix, in addition to helping fight resonant feedback.

11.13: This mixer's 100Hz lo-cut switch can fight onstage rumble and resonant feedback.

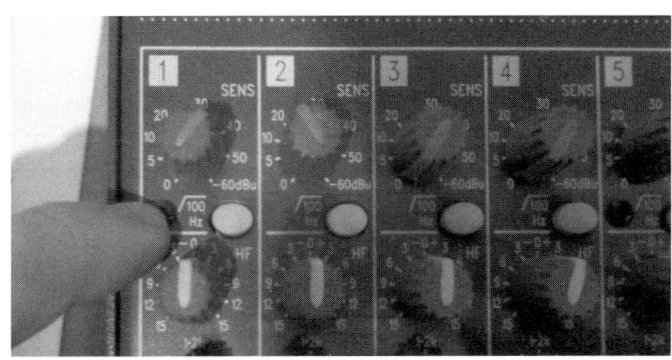

Signal Processors

Signal processors can help you adjust and refine a channel's sound. On a digital mixer and on some analog boards, such processors may be built in. Otherwise, you'll have to connect them to inserts or effects returns. Once you know they're receiving and sending signal, you can set aside the more detailed tweaking until after you've gotten a sense of the band's overall sound (see Chapter 12).

CREATING MONITOR MIXES

What makes a good monitor mix? The answer is largely up to the musicians onstage. Each player should be able to hear everything well enough to perform at his or her best. As long as the monitor mix isn't so loud that it bleeds into the stage mics, you should try to accomodate them. In an ideal world, every player onstage would have a customized monitor mix. In real life, they sometimes have to share. Communicate as clearly as possible about the

Soundcheck 101 for Musicians
Pro Tip by Charlie Lagond, Lagond Music School

The most important thing to remember is that live sound engineers don't like surprises, such as the musician playing softly during a soundcheck and then playing twice as loud during the show. Here are some tips and guidelines to follow to help the person working the mixer make you sound great.

■ When sound engineers begin a soundcheck with a band, they must first check each instrument that will be using either a microphone or direct box. Wait your turn in silence so they can focus on each instrument and communicate with each musician.

■ Make sure that the microphone is in the best place for you to play or sing into its "sweet spot" (the area where the mic's pickup is most sensitive). If the mic is unfamiliar to you, ask the engineer.

■ Demonstrate the loudest and the softest volume that you will use while performing. A pro will usually ask this of the performer, but a rookie might not.

■ Once you have established your softest and loudest levels, do not change them without informing the sound engineer! This applies to every musician (how loudly you hit the drums, set your amp, blow your horn, or sing).

■ If you do not have stage monitors, your position onstage will greatly affect how well you hear yourself and the rest of the band. Singers and horn players: If you are having difficulty hearing yourself, make sure you aren't too close to the crash cymbal or directly in front of the guitar amp.

■ Stage monitors can be set to have one or multiple mixes, depending on the mixing board and how much time the engineer has to set things up. Explain what you need on the monitors clearly and politely, and be patient while it's being set up.

■ Most stage monitors are directional, meaning that you have to stand in front of the speaker—or at least close to it—to hear it. This prevents feedback.

■ Anything that amplifies natural acoustic sound can change or color the original sound. If you find that you do not like the way you sound coming out of the P.A. speaker or monitor, the engineer can usually make adjustments by boosting or diminishing certain frequencies with the EQ.

■ If the sound is too dry—if it sounds like you're playing into a pillow—ask for some reverb to give it some ambience. But be careful not to bury yourself in the effect.

■ Don't move mics onstage without checking with the sound engineer. If you put the mic too close to a monitor or P.A. speaker, it'll create horrible and very loud feedback. And never cup or cover the active end of a mic. This can also create feedback.

■ **Important:** The sound onstage can be extremely different from the sound in the house that the audience hears. You must trust your sound engineer. If you're doing your own sound, walk around the room and listen while the rest of the band plays.

■ **Super Important:** Be nice to the sound engineer.

options and compromises in front of them. Start by sending the player's channel into the monitor mix and asking the performer if he or she can hear it in their monitor. Ask the performer if he or she wants reverb or other effects in the monitors. In some cases, the reverb will be used in the monitors, but not in the house system.

If you're onstage, be as specific as possible about what you need. Don't be afraid to speak up if you can't hear something or if you hear buzz or hum. Be polite about it; the sound engineer may be working on something else at the time. If you've got in-ear monitors, make sure they're not too loud. If something hurts your ears, let the mixer know.

Finishing Soundcheck

Once you have a basic level for each player and singer, get the group to play together. Here's where you'll set the balance among the players and make sure that everyone has what they need in their monitor mixes. If the group is going to make instrumental switches, check them as much as possible. Make sure the levels are still good following each switch—every channel should sound clean and undistorted.

Setting the balance can take some time and some practice. The bass should be solid, but not boomy. The vocals should be clear. The drumset should stand out, but not overpower everything else. Effects should enhance the sound, but not bury it.

If you have time, run through one or two songs from beginning to end. Have the players take solos and check their effects, using whatever gain boosts they plan to use onstage. Get as accurate a picture as you can. At various points in the soundcheck, move around the room. Don't be afraid to ask anyone in earshot what they think. If you're playing onstage, step off and go out into the room and listen, if possible.

Finally, ask for a set list and make note of where changes in instruments, etc., may happen. If you're happy with the sound and the musicians are happy with their monitor mix, take note of every setting—from the mixer to the amps to the instruments. This is especially important if the act will be sharing the stage and either moving its gear or letting others use it. Turn the mains back down, start up the house music, and get ready for the show.

Next Stages

Once your soundcheck is done, you're ready for the performance. The next chapter will offer a few specific tips for setting up individual sounds and mixes.

Chapter 12

MAKING
YOUR MIX

Now that everything is set up, you're ready to fine-tune your mix. Mixing for live performance is tricky because there are so many variables involved. How much of the sound will be going through the house system? How much will be coming from individual players' own rigs? How do you adjust the mix to make them all work together?

So far, we've been focusing on technical matters. If you're reading this book in chapter order, you should now be able to identify the various components in a sound system: mixer, amplifiers, speakers, microphones, effects, cables, connectors, and accessories. You should also have a basic understanding of how these pieces work together, how to set up and connect the equipment, how to position microphones, plug in instruments, get basic levels, and set up a basic house mix and monitor mix.

This knowledge should help you set up and use your own gear—and have an easier time working with a venue's sound system and sound engineer. Most musicians encounter both situations at various points in their performing lives.

Ultimately, your goal is to deliver a good-sounding performance, whether it's a solo show at a coffeehouse or a showcase at a large outdoor venue. And while a sound reinforcement system can help, it's only there to *reinforce* sound you're already producing. In other words, it can make that sound louder and deliver it clearly, but if your instruments and voices don't sound good, the P.A. system can't help you. It'll just make a bad sound louder.

In this chapter, we're going to offer some sound tips for individual players, then talk about how ensembles can practice "sounding good" the same way you practice other techniques. Finally, we'll talk about ways to use your equipment to create a pleasing mix.

Learn Your Show!

Okay, this might seem like the ultimate in obvious points. I wish that were true. But in the real world, musicians go onstage with half-learned songs, forgotten lyrics, etc. Sometimes there's no time to practice every element of a piece, but when there is, do it.

Think of the show as one entity, and then break it down into smaller pieces.

THE SET LIST

Go over the set list carefully: Which songs do you need to practice most? Look at transitions between songs and think about *everything* you need to do to move from one to the next. Are you switching instruments? Stepping up to the mic? Adjusting an effects device? Write it all down and put these notes where you can see them onstage (Figure 12.1).

12.1: Taping a detailed set list where you can see it onstage allows you to keep track of all transitions.

MUSIC AND WORDS

How well do you really know each song? If possible, try to memorize your parts. You can still write things down for backup, but memorizing the music allows you to focus your attention on the audience instead of a chart. Know the chord changes, even if you're not the one playing the chords. Develop a sense of the tempo, flow, and dynamics of each piece. Decide where you can improvise and where you need to stick to what's written.

Does part of a particular song trip you up? Study it and any other tricky passages until you're confident you can execute them. If a part is just too hard, adapt it to something easier for the show. Don't let one tough section take away from your whole performance.

Whether you're singing or not, learn the words to all the songs in the set. Knowing the meaning will influence your performance. If you're singing, can you sing each phrase clearly? Or do you mumble your way through some of them? Practice the song by speaking as well as singing the lyrics. Print lyric sheets for every song in your set, and put them in order. Keep them with you, but try not to use them onstage. Finally, make sure the song is in a good range for your voice. If it's a strain, don't force things. Change keys to a more comfortable range.

GET IN THE GROOVE

Great timing goes deeper than merely knowing how to play at a steady tempo. A great groove flows and connects the players to each other—and to the audience. When a bass drum and bass guitar are perfectly in sync and move together, the audience can feel it. It actually affects the sound—suddenly there's one "instrument" that combines the thunderous attack of the kick with the sustain and power of the bass.

So work on your timing and listen to cues from your bandmates. Sometimes, you want to hit every accent together. Other times, you might want to develop a groove where the instruments play off and in between one another. Either way, always listen! Don't just play your own part on autopilot.

Some styles of music need perfectly steady tempo. If you're playing along with prerecorded parts, sequences, loops, etc., a metronome or click track can help you come in at (and maintain) the right tempo. This can feed one player's monitors or the whole band, but it should never go to the audience. Some drummers use headphones for the click track so that there's no chance of it being picked up from the monitors.

START AND END TOGETHER

Avoid loose ends during your set. A good, strong intro will grab the audience's attention. If a song starts with just one player, everyone else should be quiet and wait until their part comes in. Don't noodle on the guitar during the drum intro!

Live shows are a great chance to show off your big endings, where everyone builds the last note to a crescendo. That's cool. But try to have a few songs that end with a quick stop. Have others fade out. Be creative. Not only does that make you sound tight; it also makes those big endings more dramatic. Eye contact and hand signals can help you end together. Watch each other, and decide ahead of time who will signal for the last note.

Practice with the Gear You Plan to Use

The best way to perfect your sound is to practice and rehearse with your performance rig every day. Spend time with the controls. Record yourself and listen for what works and what doesn't. I know it isn't always possible to rehearse with your complete rig. If you rent time in a commercial rehearsal room, for example, you might end up using the amps or drums they supply. That just means you'll have to take extra time at home to go work with your own gear.

MATCHING SOUNDS TO SONGS

Look at every song in your set and decide what kind of sound you'll use for your backing and solo parts. Are you going to switch channels on an amplifier? Use an echo effect? Change tunings? Switch from drumsticks to mallets or brushes? Use a horn mute? Change patches on a synthesizer? Run through in-song sound changes as you practice your parts. Make sure you have your moves down and that these technical details don't interfere with your ability to sing and play. Look beyond your own part. If a bandmate has to move from behind the keyboard to the front of the stage, think about where you'll be and what you'll do to keep the performance moving.

SET YOUR SOUND UP AHEAD OF TIME

Taking some time to prepare your instruments, amp, and effects settings *before* the show can save a lot of time and prevent a lot of problems at the gig.

Put Your Effects in Order

If you use effects, figure out their order of placement and test your setup. Usually, any effects that add gain (such as overdrive, distortion, compression, EQ, wah pedals, etc.) should go early in the signal chain. Spatial effects like chorus, delay, and reverb usually go later. The idea is that you create your core tone, then color it with the spatial effects.

If your amplifier has an effects loop, put gain effects between the instrument and the amp's main input. The spatial effects can then go in the loop—though some players may prefer these before the input as well.

Digital Presets

If you've got digital equipment—such as keyboards, electronic drums, preamps, effects, etc.—save presets for every song in the show. You can even store them in the order they'll be needed during the set. Presets can save time, but they shouldn't box you into a sound that may not work at a particular venue. For example, if a synth preset includes echo or reverb, try to set it up so you can adjust or mute these effects at the gig if need be. You might even create a bank of presets that are "dry" just for use in live rooms.

Marking Analog Settings

You can still create "presets" on gear that has old-fashioned knobs. You can either use a thin piece of tape to mark the knob's position, or run a strip of tape over the top of an amp and label the settings that way. You may still adjust your sound when you get onstage, but at least you'll have a starting point that was working before you left for the show. If you change anything during soundcheck, mark the new settings so you can return to them easily when you step back onstage (Figure 12.2).

12.2: A crew member labels amplifier settings during soundcheck.

Organize Your Controllers

Do you use foot-operated effects, footswitches, a volume pedal, or other controllers? Get them organized so that you don't have to set your whole rig up from scratch every time you perform. Pedalboards are great for consolidating stompboxes. If your amp has a footswitch, make sure there's room for it on or near the pedalboard, too.

Show Up Ready

The day or so before the show, make sure that all your gear is in working order. Chapter 14 explains how to make a detailed checklist. For now, let's look at the big picture. Your instrument, amps, and effects should be functioning before you leave for the gig. Never walk onstage with a piece of equipment unless you're absolutely sure it works.

Bad cables are responsible for more botched shows than anything else. You're not going to sound good through an amp or P.A. with a cable that crackles, causes ground hum, or delivers a weak signal. While you're checking cables, make sure you have a working power cable for every piece of gear, as well as any MIDI or data cables you need.

STRINGS, REEDS, AND DRUMHEADS

Having old or worn strings, reeds, and drumheads can really kill the sound of your instrument and can also make it hard to stay in tune. Some pros change their strings and drumheads every show. You don't need to do that, but you should at least do so regularly and have spares handy!

Get In Tune

Want to sound good? Play in tune. With the tuning aids available these days, there's really no excuse for any individual or ensemble to play out of tune. Tune during soundcheck (Figure 12.3), and check your tuning before you take the stage. Check tuning throughout the show. If you're out of tune and have the opportunity to do so, correct it.

12.3: A guitarist uses a pedalboard tuner during soundcheck before an outdoor show.

TUNING TOGETHER

Most electronic tuners (as well as digital instruments like samplers and synthesizers) are factory calibrated so that A=440 Hz. But this reference frequency can be changed. So make sure that everyone in the band has their instrument or tuner set to the same reference. (Some groups like to share one tuner to avoid inconsistencies; that's fine before the show, but not very useful during it.) If you're tuning to a piano that's not at A=440, see if you can recalibrate your tuners to match its A.

DRUM TUNING AND RESONANCE

Although drums don't tune to pitch the same way as strings and winds, a well-tuned drum set will always sound better than one where the drums are just set to random pitches. Always bring a key with you to the show and make sure the drumheads are in tune—they can stretch in the journey to the show. Overly resonant drums can muddy up the whole band's sound. Solve this by placing blankets and/or pillows in the kick drum; taping foam, folded cloth, or paper towels to the snare and tom heads; or using commercial mutes and dampers. Also: Turn off the snares on your snare drum when it's not in use! Vibrations from other sources rattle the snares, which can make the whole room sound like it's buzzing.

Get an Ensemble Sound

If you work on your personal sound at home, you can focus some of your rehearsal time on getting a blend with the band. Here's where you have to think about how your own sound fits with the other sounds around you.

LISTEN TO EACH OTHER AND MIX YOURSELVES

Great ensembles sound good together because every member knows when and how to adjust his or her own level. More important, every member is always listening to the other players. If your own sound stands out, you're probably too loud. Good listening habits will help you play in time and in tune. They also help you perform well when you can't hear yourself as clearly as you'd like.

Build from the Bottom

A good rhythm section is the backbone of any ensemble. Make sure the bass and drums are working together effectively. How's the timing? How's the balance between the bass and the drumset? How well do their respective parts work together?

Create Complementary Parts

How well do your individual parts fit together as a whole? Do they enhance one another, or do they fight for the same space? Let's say you have two guitars playing a chord progression at the same time. Playing in unison might work well. But if that sounds murky, maybe one can play an octave higher, use different chord inversions, or switch from block chords to arpeggios.

The more instruments in your ensemble, the more conscious you need to be of how the parts fit together. If a horn is soloing in the upper midrange, make room by getting the guitars and pianos out of the way. If the keyboard is playing a really low bass synth part, have the bass player move up an octave and double it. If the bass is soloing, keep the other instruments out of its range. Sometimes, the answer is to stop playing entirely. Try having some instruments rest during one or more verses, then come in strong on the choruses. Don't be afraid to *not* play if that sounds better!

Tones

Blending tones is also important. In an acoustic ensemble, the natural properties of the instruments will create the blend; your technique will really determine how good your tone sounds. In an amplified group, there's more opportunity to change your own sound to support or contrast someone else's.

Let's say a guitarist is playing with a brass section. A thick, overdriven sound will have similar qualities to the brass; a thin, clean sound will contrast against the brass. Try both and see what's better. If you do decide that a "wall of sound" is what you want, EQ the various parts a little differently so that they don't fight one another.

Electronic keyboards can cover an even wider range of timbres. If you're playing over guitar power chords, you might use a more percussive, staccato sound (like piano) instead of a thick synth lead to stand out. Sustaining pads can support the other instruments, even when they don't stand out on their own.

PLAY AND SING DYNAMICALLY

Sound that's relentlessly loud can be exhausting. Sound that's *always* quiet can be boring. Learn to live in the middle, and to get quieter and louder at the appropriate times. You can control onstage dynamics using performance techniques, audio equipment, or a combination of both. Master this art and you'll always have more control over your own mix.

How Much Solo Boost?

When it's your turn to solo, step forward and get closer to the mic. But when you're done, move back to your original position. Sometimes a subtle change in dynamics works better than a radical jump. One classic problem is the soloist who needs to be twice as loud as all the backing musicians combined. Yeah, you'll be heard. But you won't have the support you need to sound good—and every mistake you make will really stand out. If you can't hear yourself, move your amp or change your position relative to the monitor before turning up. If possible, ask for more "you" in the monitors. If that doesn't solve the problem, see if you can get everyone else to turn down.

Watch Your Levels

When it comes to volume, live shows can be like one of those block-building games where everyone takes turns building the thing higher until it eventually gets too tall and collapses. No singer should have to scream to be heard over the band—no matter the venue or style of music. It's hard to keep your volume under control, but you should try anyway. If your ears are ringing after practice, you're too loud.

Practice and rehearse as quietly as you can. And if one player is too loud, ask him or her to turn down before turning up your own amp. If you can't hear yourself over the drums, ask the drummer to play a little quieter. Some can do that, some can't—those in the second group should learn how to sound powerful without slamming the kit.

Playing at a reasonable volume not only protects your ears (see Chapter 13). It also enables you to hear more subtle qualities of your own instrument (and hear your bandmates more clearly). I'm not suggesting you play electric guitar and bass at acoustic guitar levels. But you don't have to be too much louder than conversational levels to get a good tone. Turning up the amp to painful levels actually masks bad tone more than it enhances good sound. Don't worry that you're not loud onstage. The P.A. will be there to support you. If you're too loud, there's nothing the engineer can do to save your sound.

Adjusting Sound with the Mixer

A mixer can really enhance the sound coming from the stage. Unfortunately, it's impossible to predict what settings will be correct for any given instrument in any given room. The EQ that works for one acoustic guitar might make another sound boxy or thin.

You'll be doing most of the basic settings during soundcheck (see Chapter 11), but be ready to make a lot of adjustments during the first couple of songs in every set. The musicians will be pumped, maybe even a little nervous, and that will translate to louder sounds, and possibly different tones. Try to anticipate problem areas ahead of time. If you're mixing yourself from the stage, don't be afraid to take a moment after the first song to ask the audience how it sounds. If they say "the bass is too loud," or "can't hear the vocals," make the necessary adjustments before the next song.

Rather than recommend specific settings, I'll go through the controls available to you and explain how they might be used.

ADJUSTING SOUND ON A SINGLE CHANNEL

When you plug a microphone or instrument into the input of a mixer channel, you're sending it down an independent path. At the end of that path, it will join any signals flowing through the other channels.

On a typical mixer, every input channel has the same set of controls (Figure 12.4). The most basic are the input gain, which sets the level of the signal coming into the channel, and the fader, which sets the level of the signal as it leaves the channel and heads to the master outputs. In between, you'll usually find an equalizer, auxiliary sends, a pan control, a mute button, and a solo button. We explained these in Chapter 2.

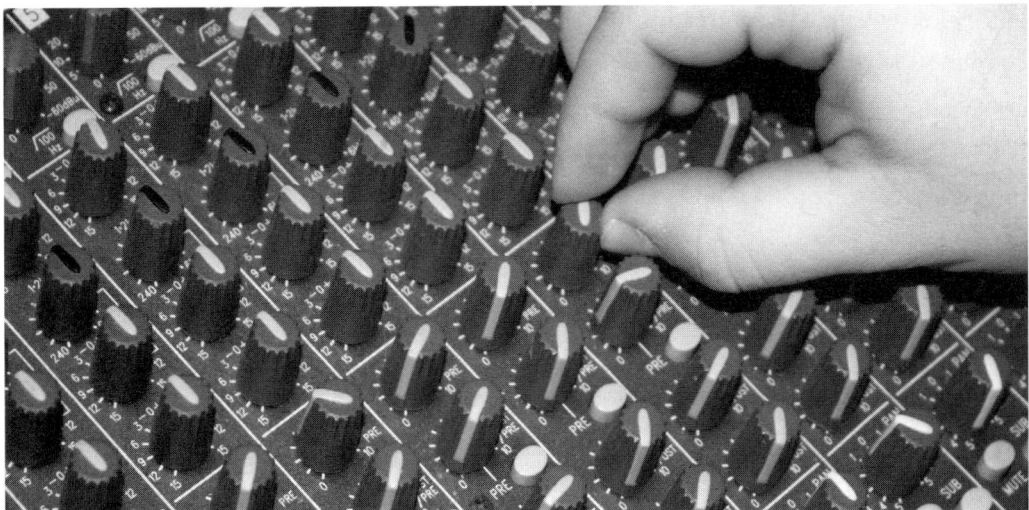

12.4: Individual channels can fine-tune the sound of each mic or instrument plugged into the mixer.

Setting Levels

We've mentioned this a few times in previous chapters, but it bears repeating. The input gain of every channel is crucial to its sound. The signal should be strong enough that you don't have to overboost it with the fader, but not so strong that it causes the channel to overload and sound distorted. The best setting depends on the level coming from the source. If a mic has a low level, you'll need to boost the input gain. If it has a high level, you'll need to lower it. Here are some tips:

Singers: Have them stand where they plan to sing, but ask them to "eat the mic" as well. Have them shout a couple of times and check the level. Mark a normal level and a hot level for the input control.

Drums and percussion: If you're using multiple microphones on a drum set, be ready for the balance to change if the drummer gets louder. In a quiet section, there may be very little bleed among the mics, but if the drummer plays harder and louder, more mics will be picking up each drum. This can cause phase problems. Don't be afraid to turn off some of the mics if necessary (you should do this with the channel mutes, rather than at the input stage). Kick drum and overhead mics are all you need for the most intense passages.

Strings and winds on mics: Whether you're working with strings or winds, the sound of an instrument will change as the player moves closer and farther from the mics. If someone is getting lost, use the monitor talkback system to ask them to step closer. If they're causing distortion when they step forward, turn down the input gain on their channel.

Pickups and electronic instruments: If a pickup connected by D.I. sounds distorted, back off on the input level. If that doesn't solve the problem, flip the pad switch on the D.I. box or ask the player to turn his or her preamp down.

Controlling Tone

You'll find a lot of variation in the EQ sections of sound mixers. Some offer only very basic two-band shelving EQs—low and high tone controls fixed to two preset frequencies. Others combine shelving bands with more complex parametric EQs that can be tuned to operate on specific frequencies and various bandwidths. Perhaps the most common arrangement will be two shelving EQs along with a semi-parametric EQ (which lets you change the frequency, but has a fixed bandwidth), as shown in Figure 12.5. You can adjust a parametric or semi-parametric EQ by sweeping the frequency control until you hear the part of the sound you want to adjust (Figure 12.6). Then use the EQ's gain knob to boost or cut that frequency.

Many mixers also have push-button filters, which you can use to automatically reduce preset frequency ranges. For example, a lo-cut filter (also known as a hi-pass filter because of the way it lets high frequencies pass through) can reduce bass rumble. This can be especially effective on stage mics.

12.5: An equalizer with low and high shelving bands and a semi-parametric midrange

12.6: This semi-parametric EQ's center frequency can sweep between 240 Hz and 6kHz.

Using EQ in a live setting can be tricky. Each room will react to different frequencies in its own way. One of the ways in which professional sound engineers really earn their money is in knowing how to use equalizers. My advice: Be conservative when using an EQ as a boost, especially when you're working with bass and vocals. Increasing gain in a channel can cause distortion and can lead to more resonant or microphonic feedback.

Remember also that the EQ you use on one channel will have to sound good when you listen with everything playing. If you boost low midrange on a guitar amp channel, you may find that doing so makes you boost lows on a bass guitar channel. You might be better off cutting lower midrange on the guitar, making room for the bass to come through.

One common problem is when a single note seems to resonate more than those around it. For example, the low A on a guitar might shake everything in the room. When that happens, have the musician play that one note while you sweep the frequency on a parametric EQ, then adjust the level. (If your mixer's EQ is limited, you can add an outboard EQ or two to your rack. These can be inserted into individual channels as needed.)

The more you know about pitches and frequencies, the easier it will be to find the right position for the EQ. A chart in Appendix 2 (there's a printable verison on the accompanying CD-ROM) shows the relationship between pitches and the fundamental frequencies they produce.

Bringing the Channel into the Mix

Two main controls set the output level of each channel. The fader sets its final loudness; the mute switch turns it on and off. A third control, pan, determines where the channel will go in the stereo sound field. Some mixers also have assign controls that determine whether a channel is going to the main mix or any submixes that are available.

Unlike the input gain control, adjusting the fader should not alter the basic character of the channel. Assuming its input gain is set properly, you should be able to use the fader to both raise and lower the channel's level. You shouldn't have to pin a fader to its highest setting, unless you're doing so for just a moment to feature a sound. Generally, the faders should sit somewhere in the middle, so that you can raise or lower a channel as needed.

Muting Channels

If a player is about to unplug an instrument and switch to another, mute the channel first! Unplugging and replugging an instrument can cause a loud pop in the monitors. The mute controls let you turn channels off when you don't need them in the mix.

Use the mute control when an instrument plugged into the mixer is not being played, when a microphone is inactive but still open onstage, or if an instrument is already loud enough onstage that it doesn't need to go through the mixer.

Adding Effects with Inserts

As we explained in Chapter 2, inserts affect individual channels. The insert point is usually very early in the channel's signal path, so any signal processors at the insert can affect the gain going through the channel. This is important. If you boost a signal with an insert effect, its gain will be higher as it heads towards the outputs. This can cause distortion, so use caution. Likewise, if you cut sound with an insert effect, the sound will be cut elsewhere in the channel.

You can use inserts for any kind of effect, but the most common uses in live sound would be to add EQ or to insert a gate and/or compressor/limiter. (You'll often find these on a single unit.) Gates and compressors help you control dynamics, and can be especially useful if there are sudden peaks coming through a channel. We explained how these work in Chapter 5.

Let's say the lead singer is also the lead guitarist. He steps away from the mic to play lead fills throughout the song, which causes bleed from other instruments to come into his mic. After playing a solo, he sometimes crowds the mic and yells, which causes the channel to distort the entire mix. Figure 12.7 shows the noise and the overload.

Inserting a gate and compressor/limiter into his channel can reduce both noise and distortion. Figure 12.8 shows the same channel with the gate and compressor/limiter active. The gate silences the channel when the guitarist steps away. When he's back on mic, the audio comes through—but when he yells, the compressor kicks in to reduce the level and stop the distortion. This makes his overall level more consistent. Compressors can also give instruments more punch in the mix. They're especially effective on bass and drums.

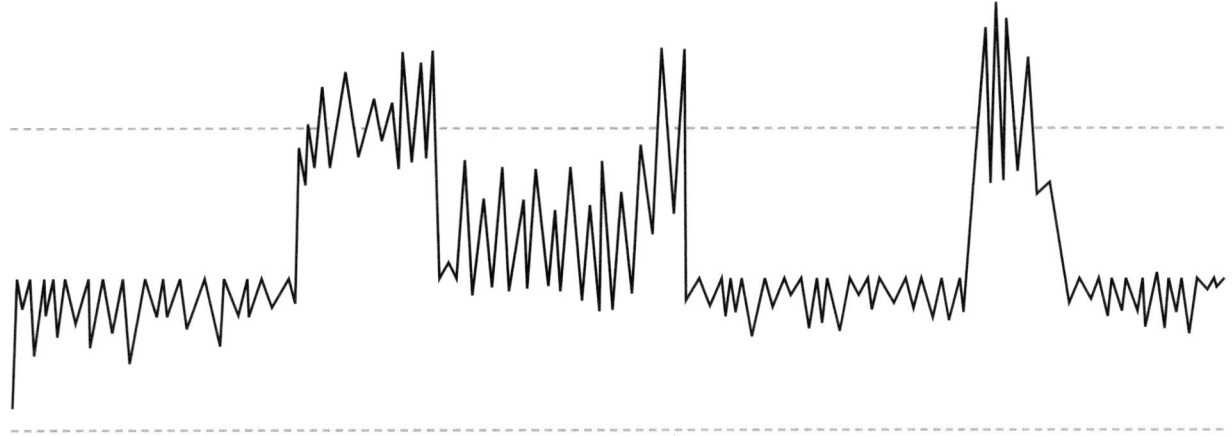

12.7: A channel with noise and wide dynamics

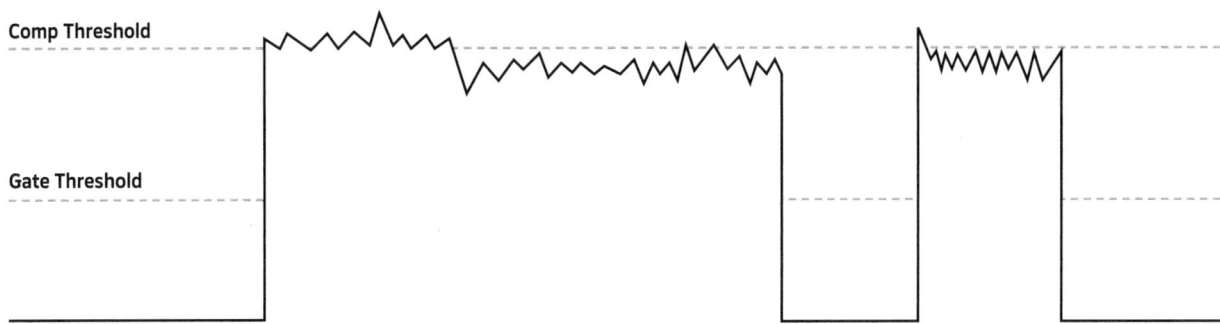

12.8: The same channel with a gate and compressor active

Adding Effects With Sends

Sends can be used for alternative mixes—such as monitor mixers—and to add effects. Adding effects with sends offers you precise control over their blend. More important, it doesn't interrupt the core sound flowing through a channel. The sends usually come after the inserts and EQs in the signal chain, so the signal going to the send has already been shaped by those devices.

The most common effects to connect to sends are reverb, delay, and chorus. If your mixer has more than one effects send, you can use one for reverb, the other for delay, etc. Usually, you'll set the effect's internal mix to 100% wet and connect its output to the auxiliary return on your mixer. (For example, if aux send 1 goes to the reverb processor, the reverb processor's output can plug into aux return 1.)

However, this doesn't have to be the case. You can use free input channels as effects returns if you want—which allows you to send the effect to a monitor mix using another aux send. I know that can seem confusing, but it's important to realize that mixers are flexible and that this flexibility gives you lots of options when setting up your sound. Figures 12.9 and 12.10 show two ways to return an effect from the same aux send to your mixer.

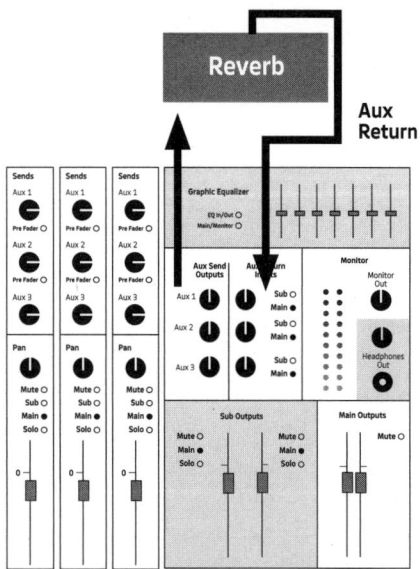

12.9: Aux send 1 is feeding a reverb. The reverb's output is mixed in with aux return 1.

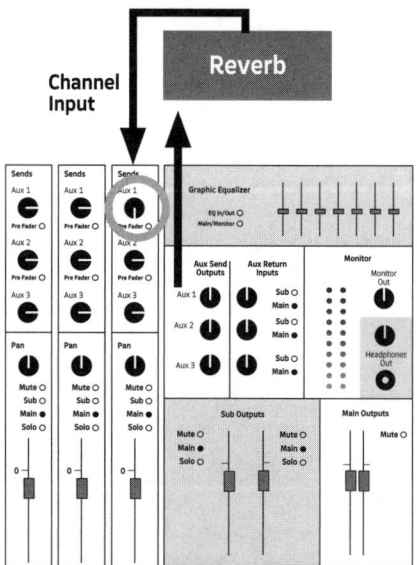

12.10: Aux send 1 feeds the reverb, but here the reverb is brought into the mix on an input channel. Note that the channel's aux 1 send is turned off to prevent feedback.

Ambient Effects in Live Venues

Reverb and echo are designed to give the listener a sense of ambience. In fact, reverb was developed as a way to imitate the sound of a concert hall in a recording studio. If you're in a room that's already pretty "live"—meaning that it has plenty of echo—adding reverb can make your mix sound muddy. Echo can also get lost in the ambience of the room.

Always listen to the mix with these effects both off and on. Try pulling back the return level on each effect until you feel that you're really missing it. Another option is to reduce the decay of the reverb—also known as the reverb time. A shorter reverb time can work better in a room where everything is sounding muddy.

You may also notice that a venue's character can change as it fills with people. Bodies and clothes absorb sound, and this will reduce the reflections in the room that cause reverberation. Once the place fills up, you may want to add some reverb back into your mix.

Using Solo to Check A Channel

If you need to make adjustments or track down a problem while the musicians perform, listen with headphones and use the solo control to isolate the channel in your headphones mix. This only works when solo is set to "monitor"—you don't want the house to hear the channel soloed! On some mixers, solo can be set to *pre-fader listen* (PFL) or *after-fader listen* (AFL). By using PFL, you can check a channel even if it's muted in the main mix.

ADJUSTING THE FINAL MIX

As the sound comes together in the individual channels, use the faders to blend them and the master outputs to set the level of the house mix feeding the power amps and speakers (Figure 12.11). There's no way to predict what settings will produce the best balance. Remember that you'll be combining your P.A. mix with the sound of the amps and instruments onstage. Perhaps you'll only use the P.A. to handle vocals and add a little low end to the kick drum during one show, and then send every instrument, even guitar amps, through the P.A. during another. When you can get a pretty consistent result in both situations, you know you're starting to master your onstage sound!

12.11: An engineer uses faders to adjust the balance between instruments during a show.

Setting Output Levels

Either way, you must watch for the proper levels at the output bus. If you have a lot of channels going at the same time, and they're all running at a pretty high output, you may end up with distortion in the master section.

Fortunately, most mixers have even more detailed meters for output than they do for the individual channel inputs. Watch the meters carefully, and try to stay in the yellow zone.

Mixing in Effects and External Sources

The auxiliary returns let you set the overall master level of effects. Let's say you like the relative level of the three vocal microphones feeding the reverb, but find that the overall reverb is too intense in the mix. Rather than reduce the level of each vocal channel, simply reduce the reverb return level. Problem solved.

Auxiliary returns can also be used as extra inputs for line-level signals. For example, you might plug a preamp or a keyboardist's submixer (Chapter 10) into a return, since all the necessary EQ and signal processing can be done at the source.

Mixing in Groups

Group buses can make it easy to control multiple channels at the same time (Figure 12.12). As we explained in Chapter 2, you can use a group to raise, lower, mute, or solo all of the drums at once. Or you can use a group to handle a bank of keyboards. To be part of the final mix, the group must be assigned to the master bus.

12.12: An engineer uses the group and master buses to control the final mix.

MASTER INSERTS

Signal processors can be used at the final output to control the overall mix. Generally, you won't apply creative effects like reverb and chorus here. The most common end-of-mix signal processors are equalizers, feedback eliminators, and compressor/limiters.

EQ

We mentioned the use of EQ on the overall mix previously. Professionals will use graphic equalizers to tune the mix to a room—for example, if the room tends to emphasize any sounds with a frequency of 120 Hz, the engineer might cut at 120 Hz to compensate.

People who are new to mixing will sometimes use the EQ to boost highs and upper midrange because they think it makes the vocals stand out. The problem with this strategy is that it also makes other instruments producing those frequencies stand out, which actually makes them compete with the vocals. Such boosting can also cause feedback.

So don't feel like you have to use a master EQ just because it's there. Try *not* to use it and then employ it only when necessary.

You can, however, use a precise notch filter to track down and reduce the level of any frequencies causing feedback. Automatic feedback eliminators do the same job by sensing and notching the offending frequencies.

Compression and Limiting

Compressors and limiters can be useful as the final effect between a mixer and power amp. A limiter can be a safety valve that tames unpredictable spikes in the overall volume—and this can save you from blowing out speaker drivers. For example, let's say a drummer accidentally slams her sticks into one of the overhead mics. That can get LOUD. If there's a limiter at the very end of the signal chain, the amp and speakers won't get the full impact of that accident.

And because it works on the entire mix, the limiter can be especially useful if there's a sudden problem on more than one channel. For example, if an acoustic guitar starts to feed back onstage, the guitar channel will obviously be affected, but the vocal mics might also pick up the feedback. A limiter will tame the output until you can mute the channels and solve the problem.

Recording Your Shows and Rehearsals

Every musician should learn the basics of recording. And while we can't deal with this complex subject in much detail, here are a few tips to help you get effective recordings from your shows.

USING A PORTABLE AUDIO OR VIDEO RECORDER

Compact digital recorders are inexpensive and effective. You should always bring one to rehearsal and use it to track your progress. Record yourself when you practice at home. Recorders with built-in mics work well when they're positioned at least a couple of feet from a loud source like amps and drums. Position them closer to sources like acoustic instruments and voices. When you're trying to capture a whole band, place the recorder in the part of the room that offers the best balance. As with any signal chain, make sure the recorder's input levels are set to match the outputs from your gear. Video recorders generally have excellent sound and also let you see yourself in action. They should be positioned like audio recorders, as long as doing so doesn't interfere with the picture (Figure 12.13).

12.13: Portable video recorders can help you see and hear yourself perform.

TAKE A FEED FROM THE BOARD

There are several ways to use a live sound mixer to record your music. If the house mix includes all instruments, vocals, and amps, a feed from the main mix can work well. You may need to use an adapter to plug it into your recorder.

If only some of the instruments are going through the mixer, you have two options.

1. You can use a sub bus that combines the main mix with some additional sources. For example, amps can be miked but kept out of the house mix and fed to a submix instead. We showed this in Chapter 2. If the amps aren't miked, use one or two condenser mics in the house to capture the overall stage sound and feed that to the sub bus and on to the recorder, with or without the other mixer input channels included.

2. Use the direct out from the channels to feed a multitrack recorder. If you're looking to actually distribute the recording (as opposed to using it for study), this is the best option if you have enough channels and tracks because it allows you to mix later.

Power Amp Settings

Unless you're using a powered mixer with a built-in amplifier, you'll be routing the mixer's outputs to power amps and monitors. The amps may be separate units, or they may be built into the speakers (which is the case with powered monitors; see Figure 12.14).

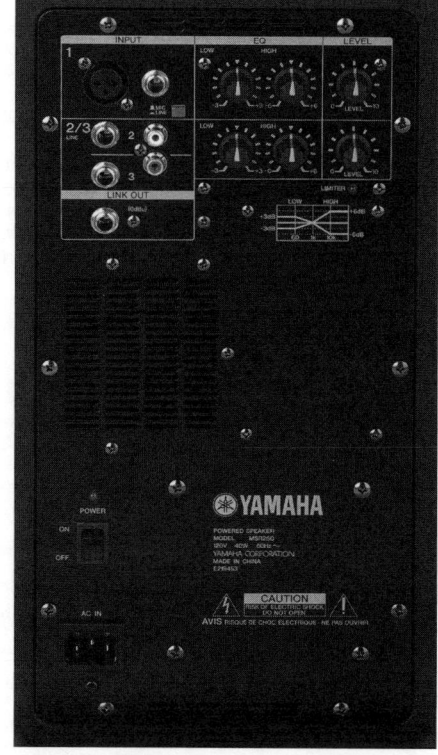

12.14: Yamaha MSR250 powered monitor with EQ

A conventional power amp usually has one volume control for each channel. These don't control the amount of power the amp generates; instead, they control the amp's input sensitivity—in a way, they're similar to the input controls on your mixer. They're set to determine what level of signal coming from the mixer produces full power from the amp. When these controls are fully clockwise, the amp has maximum headroom, and is less likely to overload. The mixer's output level should be set to match that of the amp's input. You can then control the final volume level using the mixer's master faders.

Many power amplifiers come with a built-in limiter to prevent the kinds of overloads we described earlier in this chapter. You may also be able to switch the amp from two-channel operation to a bridged mono operation. But be careful when using a high-powered amp in bridged mode—the voltage at the speaker leads could be dangerous.

Powered Monitor Controls

A powered monitor may have some additional controls that govern how the speakers work. These may include EQ as well as input sensitivity. Figure 12.14 shows the rear panel of a powered monitor that has EQ controls.

Next Stages

It takes experience and a lot of trial and error to know how to mix a show. Practice as much as possible. Mix your rehearsals like they're gigs—and record the results. Go to shows and ask to watch the sound engineer. (If you promise not to bother them and show a genuine interest in their work, you might find yourself with a very valuable mentor.) Read about sound and acoustics (you'll find a list of recommended books and websites in Appendix 3). Take an audio engineering course. Most of all, keep listening. If the music coming from the stage sounds out of balance or is too loud, make adjustments! Don't be afraid to turn things down and ask the musicians to do so. If you're one of the performers, set an example by turning down and try to get everyone else to do the same. If you have a bandmate who refuses to blend in, you might be better off getting someone else to fill that role.

In the next chapter, we'll look at hearing—which is the most important tool any musician has—and explain how you can and should protect your ears.

Chapter 13

LIVE SOUND AND HEARING SAFETY

Our ears are delicate. Loud noises damage them. It's that simple. I say this after more than 20 years of playing music at what I always considered to be "reasonable" levels. Because I never thought it was insanely loud, I hardly ever wore ear protection. As a result, I've lost quite a bit of hearing in the upper midrange.

As the Editor-in-Chief of *In Tune Monthly*, I've had the opportunity to research how music and audio affect our hearing. The following information is taken from an article I wrote in the November 2009 issue of the magazine.

Amps and Ears

When rock & roll was born in the 1950s, one of the things that made it so different was the use of amplifiers. Suddenly, a quartet with a couple of guitars, an electric bass, and a drumset could play as loudly as a big band. By the 1960s, amps got bigger and louder, and soon rock bands like the Who were standing in front of walls of speakers and setting world records for loudness.

If you met the surviving members of that group on one of their many farewell tours, you could ask them about it. But you'd probably have to speak up because they, like many musicians of their generation, suffer from severe hearing loss.

Drums and amplifiers onstage can put out a lot of sound, which can slam up against your ears like a sonic battering ram. Rocking out like the SchoolJam contestants shown in

13.1: A band rocks out in front of the amps at SchoolJam USA's 2011 Finals at in Anaheim, California.

Figure 13.1 is a lot healthier when sound levels are under control. Fortunately, you don't have to compromise your show to protect your ears.

Hearing Safety Is Cool

Years ago, people thought playing loud was cool, and nobody thought about the consequences. But today's stars—and others in the music business who make a living by listening—no longer think that blasting their ears to the point of damage is a good idea. Groups like the Audio Engineering Society, an organization for producers and sound engineers, have been encouraging musicians and audio pros to get their hearing checked and protect themselves. "The problem is not only of deep concern to our members whose livelihoods depend on their hearing," says AES Executive Director Roger Furness. "The findings of organizations such as Hearing Education and Awareness for Rockers [or H.E.A.R.] apply to virtually everyone, and can have significant impact on reducing hearing problems for future generations."

The danger doesn't just apply to rock musicians. Orchestral violinists tend to suffer loss in their left ear, while flutists suffer loss in their right. It's not that the sound is all that loud. But when you spend hour after hour exposed to it, it can do damage.

Fortunately, these pros, along with average musicians, can now use specially designed earplugs and in-ear monitors to hear their music more clearly at safe volumes. We'll look at some examples in a minute, but first, let's do a little anatomy and talk about how the ear works.

Inside the Ear

Ears are delicate and complex organs (Figure 13.2). On the outside are the *pinnae*—the flaps that stick out of your head—which help direct sound waves to the inner parts of your ear that actually send audio information to your brain.

The eardrum, also known as the *tympanum* (yes, like the drum!), separates the external ear from the middle ear. It transmits sound from the air to the *ossicles*—the three smallest bones in the human body. These send sound to the fluid-filled *cochlea*. Inside the cochlea is something called the *organ of Corti*.

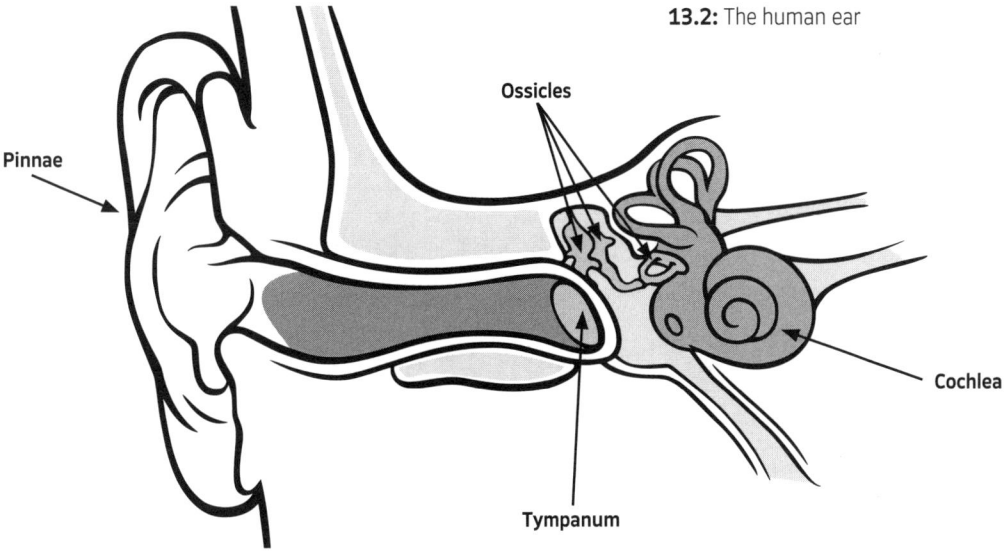

13.2: The human ear

This is where things get hairy—both literally and figuratively. The organ of Corti contains hair cells called *cilia*, and each cell can have 50 to 100 very fine hairs. These little hairs are an essential part of the hearing process. They're also the part that gets damaged when exposed to unsafe sound levels. Once cilia are damaged, they can't be restored.

HOW LOUD IS "LOUD"?

Hearing safety is an equation that adds together sound levels, measured in decibels (dB), and time. You can listen to loud sounds safely, but only for very brief periods. Rainy-day marching band rehearsal in the gym? It's safe—for about four minutes! After that, your ears are at risk unless you protect them.

Check out the chart in Figure 13.3. You can listen safely for eight hours to 85 dB of sound. But ramp it up to 111 dB, and the safe time is exactly one minute. Think it doesn't affect you? According to hearinglosshelp.com, some iPods can produce 117 dB—perfectly safe, for 15 seconds a day.

LEVEL	TIME	EXAMPLE
85 dB	8 hrs	Moderately Loud Classical Music
88 dB	4 hrs	
91 dB	2 hrs	
94 dB	1 hr	
97 dB	30 min	Loud Classical Music
100 dB	15 min	
103 dB	8 min	Loud Marching Band, Indoor
106 dB	4 min	
109 dB	2 min	Loud Rock Music
111 dB	1 min	Very Loud Rock Music
114 dB	30 sec	
117 dB	15 sec	
120 dB	8 sec	
123 dB	4 sec	
126 dB	2 sec	
129 dB	1 sec	Artillery

13.3: How long can you listen safely? The louder the audio, the shorter your ears should be exposed to it.

Keeping Your Ears Safe and Sound

So how do you prevent hearing loss? Limiting exposure is one important step, but very few musicians can restrict loud listening to only a few minutes a day. The answer is ear protection. For musicians, this falls into two basic categories:

1. Earplugs, which reduce the intensity of sound waves coming into the ear
2. In-ear monitors, which both block sounds entering the ear and provide an audio signal that the listener can control

Earplugs

Earplugs are relatively easy to use and very affordable. So why weren't musicians using them back in the hippie days? Because the earplugs they had at the time blocked sound in a way that made it hard to hear the music properly. Cover your ears with your hands and you'll get the idea; the treble and upper midrange will probably sound muffled.

Today, however, devices like Etymotic's reusable ER-20s (Figure 13.4) and custom-made Musicians Earplugs don't muffle certain frequencies. Instead, the music simply sounds qui-

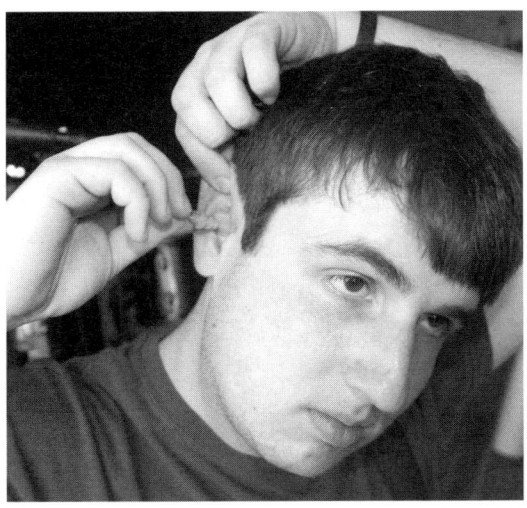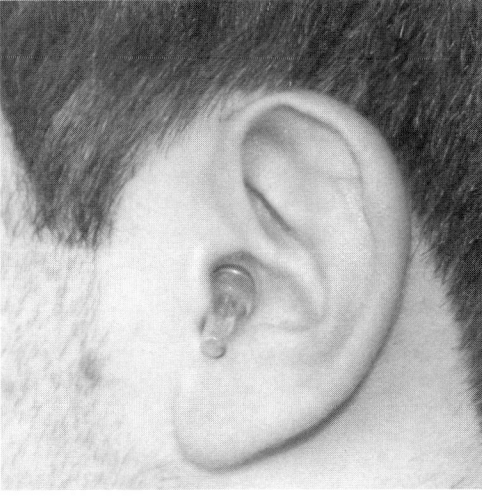

13.4: Guitarist Ben Sparks demonstrates how to put in Etymotic ER-20 earplugs.

eter. These products can reduce sound by up to 20 dB. So if you're in an environment where the sound levels are 106 dB (which is safe for about four minutes), your ears would be exposed to 86 dB (safe for five or six hours).

These kinds of plugs are ideal for electric musicians who play next to drums and loud amplifiers. (They're also great if you're in marching band.) Custom-fit earplugs, though more costly, can be even more effective at blocking sound. In either case, the plugs are reusable and fit easily into an instrument case.

In-Ear Monitors

The in-ear monitor takes the concept of ear protection one step further because, in addition to blocking the sound coming in, it also provides the listener with his or her ideal mix. Note, however, that these are not to be confused with the kind of earbuds that come with an iPod: Those don't isolate outside sounds, and can actually increase harmful levels to the ears.

Sennheiser, Shure, and other electronics companies make complete wireless systems with in-ear monitors as well. These can range in price from under $300 to more than $1,000. Examples include Sennheiser's 300IEM G2 (Figure 13.5) and Shure's PSM series—which

13.5: Sennheiser 300IEM wireless monitoring system

13.6: A bassist wears a Shure PSM earpiece with foam sleeve. (IMAGE ©SHURE INC. USED WITH PERMISSION.)

includes the affordable PSM 200.

Devices like Etymotic's ER-4 combine the sound-blocking features found in the company's earplugs with an audio monitor and can reduce external sound by 35-42 dB. They're designed to plug into any audio source, and therefore can be used with a wireless monitoring system.

These systems bundle together a transmitter, belt pack receiver, and earpieces. On Shure's PSMs, a foam sleeve over the earpiece molds to the ear and blocks outside sound (Figure 13.6). "A snare drum crack onstage can reach 120 dB," says Shure's Kevin Spiegel. "The system offers 20-40 dB isolation from the sound onstage."

You won't benefit from the in-ear monitors if you blast them in your ears. Learn to set

Setting Up In-Ear Monitors
Pro Tip by Matt Engstrom, Shure Inc.

To get the most out of in-ear monitors, it's important to put them in your ears correctly and keep your listening level under control.

Follow the manufacturer's instructions carefully. Put the earpieces in with the volume down. Make sure it feels like the earplug is firmly in place. You can test that it's blocking sound by rubbing your fingers together next to your ears. Then, slowly turn up the volume just enough so that you can hear [the audio] clearly. For further protection, the Shure devices have built-in limiters, so that a sudden loud noise won't blast the listener's eardrum.

One of the side benefits of in-ear monitoring is that when they're not straining to hear themselves, musicians are more willing to turn down their amplifiers. It's easier for vocalists to hear themselves without screaming—so they don't strain their voices. There will also be less mic feedback because there are no onstage monitors, so there's potential for overall better stage sound.

them up correctly, and get used to listening at safe levels. And if you use earplugs, practice with them in place so you're not tempted to take them out during a show because you feel like you're not hearing yourself properly.

Ultimately, you're the only person who can protect your ears. As a musician, you will be exposed to unsafe sound levels. It may happen daily. So try a pair of earplugs; get used to the way the music sounds when you wear them. Avoid blasting your ears with earbuds. Consider holding off on buying a new amp and invest in some in-ear monitors instead.

Next Stages

Get your hearing checked regularly. If your ears are ringing or you feel pain listening—stop. Hey, you wouldn't slam your fingers with a hammer and expect to play your instrument, so why do the equivalent to your ears?

Now that we've discussed your most important piece of gear, we're going to turn our attention to two of the gigging musician's biggest challenges and most important jobs: packing up and breaking down.

Chapter 14

PACKING UP, LOADING IN, SETTING UP, BREAKING DOWN

This book is supposed to be about sound. So what's all this stuff about schlepping gear? Well, as any musicians who've played outside of their own home will tell you, getting the right equipment to the right place at the right time is more than half the battle when it comes to getting great sound onstage.

In fact, other than the physical act of singing and playing your instrument, the most important part of a show's success is making sure the right gear is where it needs to be.

As an individual player, your first responsibility is for your own equipment. You should have everything you need ready and in working order at all times. As a member of an ensemble, you may also be responsible for gear shared by everyone (for example, a P.A. system). It's also a good idea to know what the other players need as individuals, and to work together to make sure everyone can do their best. After all, if the guitarist doesn't have the right cables, the drummer and singer can say, "Tough luck, pal," but their show will suffer too.

Making a Checklist

The first thing to do is make a checklist of all the essential items you need to do your job, and build from there. I'm going to start with some really obvious stuff, but before you dismiss this list as too basic, know that within those obvious things are a few items that people often neglect. I've done it; my bandmates have done it. My goal is to help you avoid joining us in the "*doh!*" club.

STRINGED INSTRUMENTS

When it comes to performing on modern stages, guitarists, bassists, and other string players have a lot to worry about. The minimal list includes:

1. Instrument (check your strings and electronics!)
2. Amp, preamp, or direct box
3. Instrument cables (or wireless transmitters)
4. Effects pedals and patch cords
5. Tuner
6. Picks (or bow and rosin), slides, whammy bars, etc.
7. Strap(s)

While not an absolute necessity, instrument stands are also pretty handy. And even if you use a power supply for effects and tuners, bring batteries.

WIND INSTRUMENTS

Wind instruments may have fewer accessories than stringed instruments, but they also need a careful check. Do you have the following?

1. Mouthpiece and reeds
2. All the keys and valves working properly
3. Accessories (such as mutes or extensions)
4. Stand(s)
5. Strap(s)
6. Tuner
7. Any electronics you use onstage

If you use your own clip-on microphone, make sure that it's working, and that any cables, transmitters, or preamps are working and have fresh batteries.

KEYBOARDS, SYNTHS, AND OTHER ELECTRONIC INSTRUMENTS

First, make sure that all the sounds and presets you need for the gig are loaded—and that you know where to find them! Be sure all controls are in working order and that the output jacks are working. Then check:

1. Power cable(s)
2. Stand(s)
3. Sustain and/or expression pedals
4. Audio cables
5. MIDI cables
6. Memory cards, discs, or any other sounds you need

You might also want to throw a set of headphones into your bag. You can use them to check and adjust your sounds in silence.

AMPLIFIERS AND EFFECTS

Like your instrument, your amplifier, effects, and cables all need to be inspected ahead of time. Check:

1. Input connections (should be tight and crackle-free)
2. Speakers and (if applicable) tubes
3. Knobs and pots (loose knobs get lost; bad pots make noise)
4. Footswitch
5. Handle, wheels, stand, and cover
6. Power cable

The power cable is especially easy to lose if it's detachable. If you're using any external cabinets, make sure you have speaker cables for all of them.

DRUMS

How many pieces are there in your drum kit? Start by accounting for those. For each drum, make sure that the skins are in good condition and that they're tight. In addition, look for:

1. All mounting hardware for both the drums and the cymbals (that includes stands, screws, etc.)
2. Kick pedal and hi-hat pedal
3. Correct beater on the kick pedal
4. Stool—check that it's working and has all of its parts
5. Mutes, blankets, pillows, or anything else you use to tame your drums
6. Every mic you need for the kit
7. Stand or clip for each one
8. Cable for every mic, as well as a couple of spares

And no matter what you do, don't forget to bring your drum key!

MICROPHONES

Doublecheck each mic's output jack, which can get loose over time. Often, this can be fixed by adjusting the small screw holding the jack in place. Make sure you have:

1. The correct clip for each mic (this is essential!)
2. Working cable
3. Stands

Even if the venue is providing the stands, it pays to bring at least one spare.

OTHER ITEMS

In addition to your working gear, make sure you have the stuff you need to sell your performance to the audience. Things like the right clothes for the show, CDs, posters or flyers, pens, and notebooks for your mailing list—they all need to be accounted for in your checklist.

Label Your Stuff

Ever go to camp? Did your mom write your name in your underwear? That may have been embarrassing, but you probably made it home with all your skivvies. Equipment gets lost very easily at gigs. It's dark. People are excited. They're focused on playing. The equipment you bring probably looks pretty similar to the gear at the venue and the stuff the other musicians have. So label it with your name and email or other contact info. Use a piece of paper secured with clear packing tape, so it won't smudge or come off by accident.

Packing Up and Loading In

Packing gear is an art. The idea is to fit everything into your car or van in a way that does no damage, keeps the gear secure, and makes loading and unloading easier.

First, assemble everything in one place (Figure 14.1), rather than dragging stuff to the

14.1: Assemble your gear in a staging area before loading equipment into your vehicle.

Loading a Van for the Gig 101
Pro Tip by Charlie Lagond, Lagond Music School

The best way to load gear, P.A. equipment, amps, and instruments into a van or car is basically to use common sense. Lots of experience also is a great help! Always remember that the ultimate goal is to get all the stuff in and get it to the gig without damaging the equipment, the vehicle, or the people traveling with the gear. Here are a few simple concepts and tips:

■ Make a checklist of everything you need so you don't forget to take anything. This is also very helpful when you are tired and packing up after a gig, so as not to forget anything you brought.

■ Pack the big and heavy equipment first; keep it as flat and low in the vehicle as possible. This will keep everything stable and safe. If you have to make a quick stop, you don't want a heavy speaker flying off the top of a pile of gear!

■ If you are good at the game of Tetris, you will probably be good at loading gear. The goal is to make everything fit so that it doesn't slide or rattle when the car moves. It usually isn't necessary to tie everything down.

■ Things like bags and drum rugs are good to use as buffers and fillers around speakers and amps and the sides of cars, and especially between equipment and windows.

■ Make sure there is no danger of anything coming loose and hitting anyone in the car.

■ It's a good idea to make a diagram of how everything is loaded so you can repeat the load at the end of the gig without too much thought.

■ Once you're on the road, if you hear something rattling and moving around, stop and make adjustments.

car piecemeal. This way, you can run through your checklist without having to go back and forth to the car. And once the car is loaded, a quick look at your staging area will tell you if you've left anything behind.

Try to think of loading into a venue as a band thing. If you play the sax and can carry your instrument in one hand, don't just walk in and sit around while the drummer makes six trips to the car for his kit. Help out! After all, you're not going to sound very good playing without drums behind you. So in a way, the drums are part of *your* rig. The same goes for the P.A. system. Everyone uses it, so everyone should help load it in and set it up.

Figure 14.2 shows a van that's been packed pretty well, courtesy of the Lagond Music School. Charlie Lagond advises that the speakers actually get packed horizontally instead of the way they're shown here, because it will make them more secure.

14.2: Pack the big and heavy equipment first and keep items as secure as possible.

Breaking Down

When the show is over, break down methodically. Look over your checklist and make sure that everything you brought is accounted for. If something's missing, tell the venue owner while you're still there instead of waiting until you get home. Not only will this make you more credible if something's been stolen, but it will also give the owner a chance to look for it among the club's stuff. Things get mislaid innocently. Even if you're tired, pack the car for the return home as carefully as you packed it for the venue. You'll be happy you did when it's time to unload.

Last Stages

Now that you know how to choose, set up, use, transport, and break down your gear, you're ready to take the stage. Remember that anyone in earshot is your audience. If you make sounding good your priority, that audience should grow and grow. Have fun up there!

Appendix 1: Glossary

Active Monitor: See Powered Monitor.

ADAT: A multichannel digital audio connection that streams eight channels of audio on one optical cable. Also known as Lightpipe and Optical.

Algorithm: A formula used by a sound module or effects device for generating sound.

Amplifier: Any device that alters the gain of a signal.

Amplitude: The loudness of an audio signal, illustrated by the height of the waveform.

Analog Input: A connection for analog audio signals.

Arpeggiator: Device for creating automatic arpeggios (chord-based musical patterns) based on notes input from a controller.

Attenuate: To decrease the level of an audio signal (i.e., make the sound quieter).

Audio Interface: Any device that can route audio to and from a computer. Types of audio interfaces include sound cards, USB interfaces, FireWire interfaces, and internal (built-in) audio.

Automation: The recording and playback of controller information, e.g., mixer settings.

Auxiliary Bus: A bus, or path, that receives signal from an auxiliary send.

Auxiliary Send: A secondary mixer channel output that routes signal to an independent destination in parallel with the channel's main output.

Balance: Stereo mixer channel control that determines a signal's position in the stereo field. See also Pan.

Balanced: An audio cable or connector with positive (hot), negative (cold), and ground connections.

Banana Plug: A type of audio connector commonly used for loudspeakers.

Bit Depth: The number of bits used to represent a single sample in a digital audio file.

Breakout Box: An external device that houses connections for an interface. Most commonly used to connect audio to a PCI card interface.

Breakout: A set of individual audio connections that emerge from one end of an audio snake.

Bridging: A means of combining two channels in a power amplifier into a single channel with more output power.

Bus: On a mixer, a common path that signals share to reach a single destination point. Examples include outputs, master outputs, internal effects sends, and groups.

Channel Strip: A set of controls for a mixer channel.

Channel: (1) A path through which audio signal travels. (2) A signal path for MIDI data.

Chord Generator: MIDI processor that can generate automatic chords based on the input of single notes.

Click or Click Track: A sound used to indicate the tempo (a.k.a., a metronome).

Clipping: Condition where the amplitude of an audio exceeds the maximum allowed level, chopping off (or clipping) the signal and causing distortion.

Co-axial: A configuration where two or more forms share a common geometric axis (e.g., a co-axial speaker driver).

Combo: An amplifier with the head and speakers in one housing.

Component System: A sound reinforcement system in which the mixer, amplifiers, and loudspeakers are individual units.

Compressor: A type of audio device that reduces dynamic range by quieting any signal exceeding a specified amount (the threshold) by a given ratio.

Control Surface: Mixer-like device that offers physical control of a software mixer.

Controller: (1) Device used to transmit MIDI messages; e.g., a MIDI keyboard. (2) MIDI message used to control parameters in real time; e.g., MIDI volume and pitch bend commands.

CPU: Central Processing Unit. This is the processor that drives a digital device.

Cross-platform: A computer application or peripheral that is compatible with more than one operating system.

Crossfade: Technique for blending two audio channels or regions to create a seamless transition.

D.I.: See Direct Box.

DAW (Digital Audio Workstation): 1) A device that combines multitrack audio recording, editing, and mixing. 2) An electronic keyboard that includes MIDI and/or audio recording capabilities and audio effects.

Decibel (dB): Unit of measure for audio level.

Digital Audio Sequencer: Software that combines multitrack digital recording, multitrack MIDI recording, mixing, and editing.

Digital Input: A connection for digital audio signals.

Direct Box: An audio device that converts a high-impedance unbalanced signal to a low-impedance balanced signal. Most commonly used to route instruments to a mixer's microphone inputs. Also known as a D.I.

Direct Monitoring: Technology that routes the inputs of the audio interface directly to its outputs, bypassing the software audio engine. Direct monitoring is used to improve monitoring latency on some native audio systems—which may be necessary if you want to use a computer in a live situation.

Download: (1) Process of transferring patches and parameter information to a digital mixer, effects device, or sound module. (2) Process for copying files onto your computer from an online source, such as a website.

Driver: (1) A loudspeaker. (2) An application that allows software and hardware to communicate.

DSP (Digital Signal Processor): Any device that's used to process digital audio; for example, a digital reverb.

DXi (DX Instrument): A software instrument that runs under the DirectX plug-in format.

Dynamics Processor: A device that can automatically control the loudness of an audio signal (for example, a compressor, limiter, expander, or gate).

Dynamics: Quality describing the loudness of a performance.

Effect: See Signal Processor.

EQ: (1) Equalization, or tone control. (2) Equalizer; a device that alters the frequency response of a signal in order to control its tone.

Expander: A dynamics processor that reduces, by varying amounts, the level of an incoming signal as it falls below a user-defined threshold.

F.O.H. (Front of House): The sound mix that the audience (house) hears.

Fade: Gradual change in audio level.

Fader: Linear control used to set signal level.

Fast Fourier Transform (FFT) Analysis: A method for analyzing the audio spectrum.

Feedback: (1) Condition where signal at an input is looped through an output and back to the input; e.g., a microphone feeding a speaker output and the speaker feeding the microphone. (2) A feature of delay and echo effects in which some of the delayed signal is fed back to the input (and back through the delay line). Increasing feedback increases the number of repeats in the delay.

FireWire (IEEE 1394): High-speed interface protocol used for connecting peripherals such as audio interfaces and digital mixers to a computer.

Frequency: The number of times a sound wave, audio signal, or modulating device travels through its complete cycle in one second. Expressed in Hertz.

Full-range: A speaker driver that reproduces low, middle, and high frequencies.

Gain Booster: A signal processor that changes the gain of a signal in order to alter its loudness, tone, or character.

Gain Stage: A part of the signal path that adds gain.

Gain: Expression of signal level.

Gauge: (1) The thickness of an audio cable. (2) The diameter of an instrument string.

Group: (1) v. To operate two or more audio channels with one control. (2) n. A bus that routes signal from several channels to a common destination.

Hard Disk Recording: Recording audio data onto a hard drive.

Headphones Amp: An amplifier designed specifically to drive headphones.

Hertz (Hz): Cycles per second.

Host: Software application from which effects plug-ins and software instruments are launched.

Hot: Describes the loudness of an audio signal; to say a signal is "too hot" means that it is too loud, and may cause distortion.

Impedance: The total opposition to the flow of electrical current or audio signal. (Ref.: *The Sound Reinforcement Handbook*.)

Inline: A type of connection that interrupts the flow of a signal path; e.g., an insert.

Input: Any connection that enables signal to enter a circuit or signal path. Examples include mixer microphone and line inputs, returns, amplifier inputs, and effects inputs.

Insert: Section in a channel strip that routes signal to an audio processor in series with the channel's signal flow.

Instrument Amplifier: Any amplifier designed for a specific instrument (e.g., guitar, bass, acoustic guitar, keyboards).

Instrument Level: The typical audio signal level produced by instrument pickups (typically -10dBu).

Interface: A device that routes signal to and from the computer; e.g., audio interface, MIDI interface.

Latch: Automation mode in which changes are recorded from the time you touch a control until the time you stop playback.

Latency: The delay between input and output of a signal as it travels through a digital audio system.

Library: A system for organizing effects and instrument patches into categories.

Lightpipe: See ADAT.

Limiting: Dynamics processing that stops any signal from exceeding a threshold.

Line Level: A typical audio signal level at the output of electronic devices (+4dBu balanced, -10dBV unbalanced).

Line Mixer: A mixer designed for line-level signals.

Link: To connect two or more devices (such as mixers and signal processors) to operate as one unit.

Loop: (1) v. To repeat. (2) n. Section of audio or MIDI that repeats within an arrangement.

Main Bus: See Master Bus.

Master Bus: The main output bus on a mixer.

Mic Level: A typical audio signal level produced by microphones (-20dBu balanced).

MIDI (Musical Instrument Digital Interface): Protocol for sending control signals between compatible devices.

MIDI Clock: MIDI synchronization signal that follows bars and beats.

MIDI Machine Control (MMC): A two-way communications message that allows one device to control the transport of another.

MIDI Port: A MIDI connection that can send or receive 16 MIDI channels.

MIDI Time Code (MTC): MIDI synchronization signal that follows hours, minutes, seconds, and frames.

Mix: To blend multiple signals for final output; e.g., mixing a multitrack arrangement to a stereo format.

Modeling: Digital technology that creates sound by emulating the performance of another device; e.g., a modeling instrument based on an analog synthesizer, a modeling signal processor based on a tube guitar amp.

Modulation: Type of audio effect that uses delay and pitch controls to alter the sound over time. Examples include chorus and flanger.

Monitor: (1) v. To listen. (2) n. Playback system used to listen to audio.

Multiband Processor: A signal processor that divides the audio signal into sections, or bands, based on frequency, and processes them independently.

Multiport MIDI Interface: A device that offers multiple independent MIDI connections, each capable of transmitting or receiving 16 MIDI channels.

Multitimbral: A MIDI instrument capable of responding to more than one MIDI channel at a time.

Multitrack: A device capable of recording and playing back more than two tracks of audio.

Mute: (1) v. To silence a signal. (2) n. A switch on a mixer or other audio device that silences the output of a channel or bus.

Native: An audio system that utilizes a computer's internal processor for signal processing.

Noise Gate: Dynamics processor that attenuates any signal that falls below a threshold.

Note: MIDI message that triggers a sound in a MIDI device.

Output: Connection or bus by which signal is sent to a destination.

Over: A signal that exceeds the amplitude limit of a digital signal path, causing distortion.

Pad: A switch that attenuates an audio signal.

Pan: Control for positioning a signal in the stereo (or surround-sound) field.

Parallel: Drawing the signal from a channel without interrupting the signal flow, as with an effects send.

Parameter: Any element in a device that can be controlled.

Patch Bay: A device that allows you to consolidate audio connections from multiple sources.

Patch Cord: A short cable used to interconnect audio devices. Uses include connecting effects on a pedalboard and connecting a mixer to an effect on a patch bay.

Patch: (1) n. A collection of settings that can be stored for later recall. (2) v. To connect two signals or devices together; e.g., patching a compressor to an insert with a patch bay.

Peak: Maximum amplitude, or loudness, of a waveform.

Phase: The relative polarity of a sound wave or audio signal.

Physical Input: A hardware input connection on an audio interface or digital mixer.

Physical Output: A hardware output connection on an audio interface or digital mixer.

Pickup Pattern: Term describing the directionality of a microphone.

Pickup: A transducer that converts string or instrument body vibrations into audio signal.

Pitch Bend: MIDI continuous controller message that's used to change the pitch of a note over a specified range.

Pitch Correction: Signal processing that conforms an audio file's pitch to a preset scale.

Playlist: A collection of tracks designated for playback.

Plug-in: Software that extends the capabilities of a host. Most commonly refers to audio effects processors and software instruments.

Polarity: Positive and negative positions of an electrical current or audio signal.

Power Amplifier (Power Amp): (1) A device that raises a signal to a voltage that can drive a set of speakers. (2) The power section in an amplifier.

Powered Mixer: A mixer with a built-in power amplifier.

Powered Monitor: A speaker enclosure with a built-in power amplifier. Also known as an Active Monitor.

Preamplifier (Preamp): A standalone device or section of a larger device that boosts an input signal's gain so it can be amplified electronically.

Program: (1) n. A memory location for storing parameters in a digital instrument, effects device, or mixer; e.g., a synthesizer patch or effects setting. (2) v. To edit or create a sound or a part on a synthesizer, sequence, or effects device.

Proximity Effect: The change in a microphone's frequency response as it moves closer to a source.

RAM: Random Access Memory.

Read: To play back mix automation data.

Real-time Processing: Signal processing that's applied to a source as the source plays.

Resolution: The number of subdivisions in a given time format. MIDI resolution measures the number of pulses per quarter note (PPQ). Audio resolution measures the number of samples per second (see Sample Rate).

Return: An input that routes signal from an external source or internal bus into the main mix; e.g., effects return.

ReWire: A software routing protocol that's used to connect audio and MIDI between compatible applications.

Roll-off: A reduction in level.

Sample Rate: The number of samples per second, expressed in Hertz.

Sample: (1) The smallest unit in a digital audio recording. At a sample rate of 44.1 kHz, there are 44,100 samples per second of audio. (2) An audio recording that can be triggered in real time by a sampler.

Sampler: An electronic instrument that can record and play back short audio files, called samples.

Send: Mixer control that routes signal to an effects bus, output, or other destination.

Sequencer: Traditionally, a data recorder that stores MIDI commands or events. Current usage also includes DAWs.

Serial: Audio connection in which a device interrupts the signal flow (see Inline).

Shield: Part of an audio cable that rejects electronic and magnetic interference.

Signal Path: The path a source signal takes on its way to a destination. Also known as signal flow.

Signal Processor: Any device that alters a signal. Most commonly refers to audio effects.

SMPTE-to-MIDI Converter: Device that translates SMPTE time code to MIDI Time Code (MTC).

SMPTE: Linear time code used by the Society of Motion Picture and Television Engineers that expresses time in hours, minutes, seconds, and frames.

Snake: A group of cables secured together in one large lead. Usually refers to audio cables with an input box on one end and a breakout of individual cables on the other.

Snapshot: A form of static automation that stores all current mixer settings for later recall.

Soft Synth: See Software Instrument.

Software Instrument: A software emulation of a physical instrument, such as a synthesizer.

Solo: (1) v. To silence all but the selected mixer channel(s), track(s), or region(s). (2) n. Mixer control used to isolate the playback of a channel.

Sound Card: A computer's audio interface.

Speakon: A type of locking audio connector commonly used for loudspeakers.

SPL: Sound Pressure Levels.

Split: (1) To send one audio signal to two separate destinations. (2) To separate a stereo signal into left and right mono channels.

Stack: An amp with a separate head and speakers.

Stand-alone: Refers to any device or application that can function independently.

Submix: A mix within a mix. See Group.

Subwoofer: An ultra-low frequency driver.

Sync: Synchronize.

Synchronize: To lock two devices so that they play back together.

Synthesizer: An instrument that produces sound by manipulating an electronic (or digital) signal. Synthesis techniques include additive synthesis, frequency modulation (FM), wavetable synthesis, digital modeling, and others.

System Exclusive (SysEx): MIDI messages used for transmitting parameter data between devices.

Template File: A data file that can be used by an application to set the basic parameters of a project. For example, a template file for a digital mixer might be used to zero the board.

Thru: (1) A connection that sends the signal from an input to an output without modifying it; e.g., the unbalanced pass-through connection on a direct box. (2) Avenue for MIDI transmission that passes through a device on its way to another.

Touch: Automation mode in which changes are recorded only while a control is being operated. When you let go of the control, the automation reverts to previously recorded automation information.

Transducer: A device that converts vibrations into electrical energy, such as a microphone or pickup.

Transformer: (1) An audio device that changes the impedance and balanced/unbalanced nature of a signal to match that of an input. (2) A device that transfers electric energy from one circuit to another.

Transpose: To change pitch or key, either through performance techniques or by electronic means.

Trim: Hardware mixer control used to set the input level on a channel.

TRS (Tip/Ring/Sleeve): A type of audio connector used for balanced and stereo connections

TS (Tip/Sleeve): A type of audio connector used for mono unbalanced connections.

Tweeter: A high-frequency speaker driver.

Unbalanced: An audio cable or connector with positive (hot) and ground connections.

Universal Editor/Librarian: Application that can edit and store MIDI System Exclusive information for a variety of devices from different manufacturers.

USB (Universal Serial Bus): Interface used for connecting controllers, digital mixers, and other devices to a computer.

Velocity: A MIDI note parameter that measures the force with which a key is struck. Velocity is usually used to control dynamics, but can be mapped to other attributes, such as tone.

Volume: Loudness of an audio signal.

VST (Virtual Studio Technology): A native audio plug-in format that's compatible with a wide variety of effects and host applications.

Waveform: A visual display of an audio file.

Woofer: A low-frequency speaker driver.

Word Clock: Digital clock format that can be used to synchronize a number of digital audio devices to one master clock.

Write: To record automation data.

Appendix 2: Frequency to Pitch

Frequency Hz	Pitch	Frequency Hz	Pitch	Frequency Hz	Pitch
16	CØ	131	C3	1047	C6
17	C#/Db	139	C#/Db	1109	C#/Db
18	D	147	D	1175	D
20	D#/Eb	156	D#/Eb	1245	D#/Eb
21	E	165	E	1319	E
22	F	175	F	1397	F
23	F#/Eb	185	F#/Eb	1475	F#/Eb
25	G	196	G	1568	G
26	G#/Ab	208	G#/Ab	1661	G#/Ab
28	A	220	A	1760	A
29	A#/Bb	233	A#/Bb	1865	A#/Bb
31	B	247	B	1976	B
33	C1	262	C4	2093	C7
35	C#/Db	278	C#/Db	2218	C#/Db
37	D	294	D	2349	D
39	D#/Eb	311	D#/Eb	2489	D#/Eb
41	E	330	E	2637	E
44	F	349	F	2794	F
46	F#/Eb	370	F#/Eb	2960	F#/Eb
49	G	392	G	3136	G
52	G#/Ab	415	G#/Ab	3322	G#/Ab
55	A	440	A	3520	A
58	A#/Bb	466	A#/Bb	3729	A#/Bb
62	B	494	B	3951	B
65	C2	523	C5	4186	C8
69	C#/Db	554	C#/Db	4435	C#/Db
73	D	587	D	4699	D
78	D#/Eb	622	D#/Eb	4978	D#/Eb
82	E	659	E	5274	E
87	F	699	F	5588	F
93	F#/Eb	740	F#/Eb	5920	F#/Eb
98	G	784	G	6272	G
104	G#/Ab	831	G#/Ab	6645	G#/Ab
110	A	880	A	7040	A
117	A#/Bb	932	A#/Bb	7459	A#/Bb
124	B	988	B	7902	B

Appendix 3: Resources

MANUFACTURERS

AKG
akg.com
Accessories, Cables, Cases, Headphones, In-ear Monitors, Microphones, Stands, Wireless

Alesis
alesis.com
Amps, Cables, Effects, Keyboards, Mixers, Monitors, Recorder, Software, Speakers

Allen & Heath
allen-heath.com
Accessories, Effects, Mixers, Software

Aphex
aphex.com
Accessories, Amps, Effects, Headphones

ART
artproaudio.com
Accessories, Amps, Effects, Microphones, Mixers

Ashly
ashly.com
Accessories, Amps, Effects, Mixers

Atlas Sound
atlassound.com
Accessories, Amplifiers, Cables, Speakers, Stands

Audio-Technica
audio-technica.com
Accessories, Headphones, Microphones, Wireless

Audix
audixusa.com
Accessories, Microphones, Speakers, Stands, Wireless

Aviom
aviom.com
Accessories, Mixers, Software

Barcus Berry
barcusberry.com
Pickups

BBE
bbesound.com
Accessories, Effects, Software

Beyerdynamic
beyerdynamic.com
Accessories, Cables, Headphones, Microphones, Stands

Blue Microphones
bluemic.com
Accessories, Cables, Microphones, Recorders

Bock Audio
bockaudiodesigns.com
Microphones

Bose
bose.com
Accessories, Amps, Headphones, Mixers, Monitors, Speakers

Cakewalk
cakewalk.com
Accessories, Effects, Mixers, Software

Chandler Limited
chandlerlimited.com
Accessories, Effects, Mixers

ClearSonic
clearsonic.com
Accessories, Isolation Booths

Coleman Audio
colemanaudio.com
Monitors

Countryman
countryman.com
Accessories, Microphones

Crown
crownaudio.com
Amps, Microphones

dbx
dbxpro.com
Accessories, Effects

Denon
denon.com
Accessories, CD/DVD Players, Headphones, Speakers

DigiTech
digitech.com
Accessories, Amps, Effects

DPA
dpamicrophones.com
Accessories, Microphones

Drawmer
drawmer.com
Accessories, Effects, Monitors

DiMarzio
dimarzio.com
Accessories, Cables, Pickups

Dean Markley
deanmarkley.com
Accessories, Amps, Pickups

E-MU
emu.com
Accessories, Digital Audio Systems, Keyboards, Software, Wireless

Earthworks
earthworksaudio.com
Accessories, Microphones, Pre-amps

Electro-Voice
electrovoice.com
Accessories, Amps, Microphones

Etymotic Research
etymotic.com
Earphones, Headsets

Eventide
eventide.com
Accessories, Effects

EMG
emgpickups.com
Accessories, Pickups

Fender Audio
fender.com
P.A. Systems, Stands

Fishman
fishman.com
Amps, Effects, P.A. Systems,
Pickups

Furman
furmansound.com
Accessories, Amps, Mixers

Galaxy
galaxyaudio.com
Monitors, Accessories,
P.A. Systems

Gator
gatorcases.com
Accessories, Cases, Stands

Groove Tubes
groovetubes.com
Accessories, Microphones, Preamps

Hear Technologies
heartechnologies.com
Accessories, Amps, Cables,
Mixers, Monitors

Hearos
hearos.com
Earplugs

Hosa
hosatech.com
Accessories, Cables

JBL
jbl.com
Accessories, Monitors,
P.A. Systems, Speakers, Stands

JDK Audio
jdkaudio.com
Effects, Preamps

Korg
korg.com
Accessories, Cases, Effects, Keyboards, Recorders, Stands

Lauten Audio
lautenaudio.com
Microphones

Levy's
levysleathers.com
Guitar Straps

Lexicon
lexicon.com
Accessories, Amps, Effects

Line 6
line6.com
Accessories, Amps, Effects,
Microphones, Wireless

Listen Technologies
listentech.com
Accessories, Cables, Headphones,
Monitors

LR Baggs
lrbaggs.com
Amps, Pickups, Preamps

M-Audio
m-audio.com
Accessories, Cables, Cases,
Effects, Microphones, Monitors,
Speakers

Mackie
mackie.com
Accessories, Cases, Mixers,
Monitors, Speakers

**Middle Atlantic
Products**
middleatlantic.com
Accessories, Panels, Shelves

Mighty Bright
mightybright.com
Accessories, Stands

Miktek
miktekaudio.com
Microphones, Preamps

Mojave Audio
mojaveaudio.com
Accessories, Microphones

Monster
monstercable.com
Accessories, Cables, Headphones

Moog
moogmusic.com
Accessories, Cases, Effects

Music Accessories
music-accessories.net
Panels, Racks, Shelves

Neumann
neumannusa.com
Accessories, Cases, Microphones,
Monitors, Stands

Omnimount
omnimount.com
Accessories, Stands

On-Stage Stands
onstagestands.com
Accessories, Stands

Peavey
peavey.com
Accessories, Amps, Effects, Microphones, Mixers, Monitors, P.A.
Systems, Speakers, Stands

Pioneer Pro DJ
pioneerdj.com
Effects, Headphones, Mixers

PreSonus
presonus.com
Accessories, Cables, Mixers,
Preamps, Software

Primacoustic
primacoustic.com
Accessories, Panels, Stands

Pro Co
procosound.com
Accessories, Cables

Provider Series
providerseries.com
Accessories, Microphones

QSC
qscaudio.com
Accessories, Amps, Cases,
Speakers

Radial
radialeng.com
Accessories, Boxes, Effects,
Preamps

Rane
rane.com
Accessories, Amps, Effects,
Mixers, Software

Raxxess
raxxess.com
Accessories, Panels, Racks,
Stands

RME
rme-audio.com
Accessories, Cables

Rode
rodemic.com
Accessories, Cables, Cases,
Microphones

Roland
rolandus.com
Accessories, Amps, Cables, Cases,
Mixers, Monitors, P.A. Systems,
Pickups, Software, Stands

Royer
royerlabs.com
Accessories, Microphones

Sabine
sabine.com
Accessories, Tuners

Sabra-Som
sabrasom.com.br
Accessories, Stands

Samson
samsontech.com
Accessories, Amps, Headphones,
Microphones, Mixers, Monitors,
Stands, Wireless

Schoeps
schoeps.de
Accessories, Microphones

sE Electronics
seelectronics.com
Microphones, Stands

Sennheiser
sennheiserusa.com
Accessories, Cables, Headphones,
Microphones, Monitors, Software,
Stands, Wireless

Shure
shure.com
Accessories, Cables, Headphones,
Microphones, Mixers, Monitors,
Stands, Wireless

SKB
skbcases.com
Accessories, Cases, Shelves

Sony
sony.com
Accessories, Cables, Effects,
Headphones, Microphones,
Mixers, Software

Sound Devices
sounddevices.com
Accessories, Cables, Cases,
Mixers, Recorders

Soundcraft
soundcraft.com
Accessories, Mixers

Studio Projects
studioprojectsusa.com
Accessories, Microphones,
Preamps, Shelves

Studio Technologies
studio-tech.com
Monitors, P.A. Systems

Switchcraft
switchcraft.com
Accessories, Adapters,
Connectors

TASCAM
tascam.com
Accessories, Cables, Effects,
Mixers, Recorders

TC Electronic
tcelectronic.com
Accessories, Amps, Effects,
Software

TC-Helicon
tc-helicon.com
Accessories, Cases, Effects

Teac
teac.com
Headphones

Ultimate Support
ultimatesupport.com
Accessories, Cases, Stands

Universal Audio
uaudio.com
Accessories, Preamps

Vocalist
vocalistpro.com
Effects

Waves
waves.com
Accessories, Effects, Software

Whirlwind
whirlwindusa.com
Accessories, Cables

Yamaha
yamaha.com
Accessories, Amps, Cases,
Headphones, Keyboards, Mixers,
Monitors, Speakers, Stands

ORGANIZATIONS

**National Association
of Music Merchants**
namm.org

**The National
Association for
Music Education**
menc.org

**Audio Engineering
Society**
aes.org

ADDITIONAL
RESOURCES

*dpamicrophones.com/en/
Microphone-University/
ApplicationGuide.aspx*

shurenotes.com

Photo Credits